Co-Treatment of Septage and Fecal Sludge in Sewage Treatment Facilities

Co-Treatment of Septage and Fecal Sludge in Sewage Treatment Facilities

A Guide for Planners and Implementers

Dorai Narayana

Published by   **IWA Publishing**
Alliance House
12 Caxton Street
London SW1H 0QS, UK
Telephone: +44 (0)20 7654 5500
Fax: +44 (0)20 7654 5555
Email: publications@iwap.co.uk
Web: www.iwapublishing.com

First published 2020
© 2020 IWA Publishing

British Library Cataloguing in Publication Data
A CIP catalogue record for this book is available from the British Library

ISBN: 9781789061260 (print)
ISBN: 9781789061277 (eBook)

Contents

1 BACKGROUND

Most developed countries use water-borne sewerage systems to manage urban sewage. Many developing countries take this as a model for the future in their own cities, and these countries too have tended to lean towards networked sewerage to solve urban sewage problems. Sewerage is a good solution which protects public health and the environment by effective containment, transport and treatment of sewage. However, it is a very expensive solution, and difficult to implement effectively. Nevertheless, many cities in the developing world already have sewage treatment facilities, and many more are planning to install such facilities.

Over the past few years, on-site treatment has been widely promoted as a process, which can be implemented quickly, to address problems in sanitation and is becoming more widely accepted. As such, treatment of the contents of on-site containments (septic tanks, vaults or pits) has become a pressing issue. While dedicated treatment facilities for this purpose have been advocated, co-treating these wastes in sewage treatment facilities is also a promising option, which many countries have implemented or are cautiously exploring. This option maximises the utilisation of city infrastructure, and is therefore advantageous. In cases where the existing sewage treatment facilities are underutilised, co-treatment presents a ready solution for managing fecal sludge (FS) and septage. In developed countries, co-treatment is practiced widely, but with very small quantities of septage in comparison to the sewage flows. This is because very large portions of the cities are extensively sewered, and on-site systems are few. The situation in developing countries is often reversed, with large parts of the cities using on-site systems.

© IWA Publishing 2020. Co-treatment of Septage and Fecal Sludge in
Sewage Treatment Facilities
Author: Dorai Narayana
doi: 10.2166/9781789061277_0001

In spite of co-treatment being a well-known practice in many countries, it remains clouded in uncertainty, especially regarding the technical advisability, and potential risks of co-treating FS or septage in sewage treatment plants (STPs). Planners and decision makers are often very apprehensive in considering co-treatment. As a result, the opportunity to better utilise available infrastructure for co-treatment of sludge is often being missed.

Meanwhile, there are also many cases where co-treatment has been tried, either successfully or otherwise, but it has not been possible to draw conclusions from these to guide the way forward. Case studies of such instances have been documented, but the cause of the success or failure often cannot be conclusively found. The situations involved are highly variable, and reliable or complete related data is often not available to enable a proper evaluation.

This guideline aims to explore some of the basic principles behind sewage treatment, and how it may be impacted by wastes from on-site containments, to try to throw some light on how co-treatment could be considered, in an incremental manner, recognising risks and mitigating them.

2 OBJECTIVES OF THIS GUIDELINE

This guideline is intended to facilitate a better understanding among planners, engineers, decision makers and technical practitioners in the following aspects:

(a) the relevant differences between FS/septage and sewage
(b) situations in which co-treatment may be considered
(c) the potential of co-treatment of septage/FS in STPs
(d) issues of concern in co-treatment, potential impacts and mitigations
(e) hand holding in a step-by-step consideration of co-treatment planning

It is hoped that with this understanding, available case studies can be better understood and potential strategies mapped out for each local situation. Some cautionary notes are also included for the practitioner.

The information included here relies on already available published material, particularly *Fecal Sludge and Septage Treatment: A guide for low- and middle-income countries* by Kevin Tayler (2018), although this book deals with fecal sludge management in general and co-treatment is not dealt with in great detail.

It must be stressed that this is not a design manual. It is meant as a guide for planners to evaluate and consider the option of co-treatment.

3 TERMINOLOGY

The term co-treatment may refer to treating different wastes together, for example liquid septage with municipal solid wastes, liquid septage or FS with sewage, or partially solid FS with sewage sludges. This guideline is concerned primarily with treating septage or FS together with sewage.

The term sewage is used to denote human excreta, mixed with wash water, flush water as well as grey water (from bathrooms, laundry, kitchen and other domestic sources).

The terms Septage and Fecal Sludge are defined as below:

- *Fecal sludge is* the material which accumulates at the bottom of a pit, tank or vault, where there is little water added, or the bulk of the water has overflowed/percolated away.
- *Septage* refers to the solids and liquids which are removed from a pit, tank or vault in a wet sanitation system, and comprises FS, the supernatant water and scum.

In this guideline, the term 'sludge' is used to denote the emptied contents of on-site system containments, and includes septage, FS, contents of container-based vaults, community toilets, mobile toilets as well as combinations of all these. Sludge produced as a STP by-product is termed 'STP sludge'.

4 WHY CO-TREATMENT

The last few years have seen a huge increase in interest in implementing fecal sludge management, in parts of the world where sewered systems are absent, or are few and cover small populations. In the absence of sewers, households will have to build on-site systems for human excreta management. This has resulted in the need to develop suitable and appropriate systems to manage sludge removed from on-site system containments. The components of such a system are shown in Figure 1.

Most solutions consist of interventions in the various parts of the value chain, from containments to emptying/transport of sludge and treatment, disposal of end products and reuse. Major focus has been on providing septage/FS treatment facilities.

Utilising existing sewerage infrastructure for treatment of sludge has gained attention, particularly in India, because of the existence of a number of STPs which are underutilised (and expected to be underutilised for the near future). This has opened up the potential of co-treatment of sludge in existing STPs. The potential includes utilising existing STPs, with or without retrofits, and new STPs being designed to co-treat sludge. The large number of new STPs being planned or existing STPs expected to be upgraded or retrofitted in next few years creates a huge opportunity for co-treatment.

Many large cities, especially in India, either already have a sewerage system with a STP, or plan to have one in the near future. Sewering existing cities completely, with all sources of sewage connected to the sewer network will be impossible, and cities will continue to have pockets relying on on-site systems. These on-site systems will then need a septage/fecal management programme, and if appropriate, the sewerage infrastructure could be utilised to co-treat the septage/FS, either on a permanent basis, or as an interim arrangement until dedicated septage/FS treatment facilities are built. It is prudent

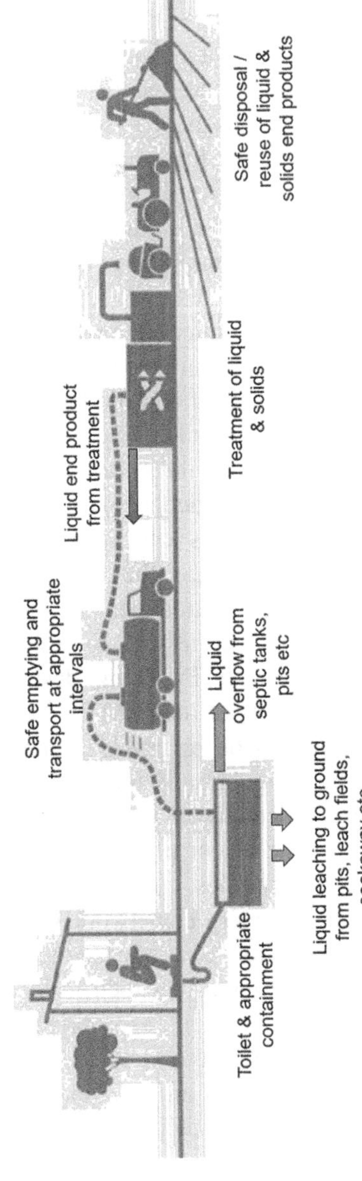

Figure 1 Fecal sludge management components. *Source*: Adapted from Bill & Melinda Gates Foundation.

to consider co-treatment as an option, to maximise available infrastructure wherever possible.

There is also opportunity for other towns in close proximity to these treatment plants for co-treatment (in practice, towns within a radius of 10–12 km of an STP could potentially bring their septage to be co-treated in the STP).

5 SEWAGE, SEPTAGE AND FECAL SLUDGE
5.1 Compatibility of sewage and septage/fecal sludge

Sewage as defined above is human excreta, mixed with wash water (where used) and flush water as well as grey water (from bathrooms, laundry, kitchen and other domestic sources). This is conveyed through a system of pipes, pumps etc to a treatment facility. It is primarily (>99%) water, and the remainder consists of organic and inorganic matter in dissolved or suspended form, nutrients and pathogens. Typically, sewage reaches the treatment facility in a matter of hours, and it is still fresh. Typical characteristics for sewage are shown in Table 1 (from various sources). As can be seen, there is some variability in the figures, which could be due to local factors such as household size, water use, etc.

When an on-site system is used, the excreta, wash water and flush water are conveyed to an on-site containment. This could be a septic tank, a pit, a vault or simple containment tank (see Figure 2). The containment could be water tight, open bottomed or porous. It may have an outlet for the supernatant to soak into the ground or flow into surface drains. It may also allow groundwater to infiltrate or surface water to backflow inside the containment. Depending on various factors such as the design of the containment, the local ground conditions and the period for which the waste remains in the containment, different processes occur inside the containment. This includes settling, consolidation, dilution and anaerobic digestion of the solids. The nature of the waste will undergo a corresponding

Table 1 Typical characteristics of domestic sewage.

Biochemical oxygen demand	150–250 mg/l
Total suspended solids	200–350 mg/l
Chemical oxygen demand	300–500 mg/l
Total Kjeldahl nitrogen	35–50 mg/l

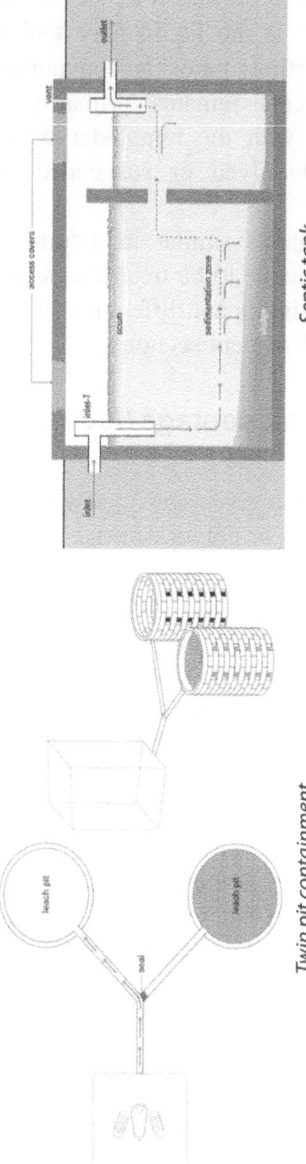

Figure 2 Examples of containment systems. *Source:* EAWAG. (2014). Compendium of sanitation systems and technologies 2nd Revised Edition. (Dübendorf, Switzerland: Swiss Federal Institute of Aquatic Science and Technology (Eawag)).

transformation, and part of the solids will settle out and undergo anaerobic digestion, while the liquid may soak away into the surrounding soil or overflow out of the containment.

However, the contents remains primarily water (>95% and typically >98%), with the remainder being organic and inorganic matter in dissolved or suspended form, nutrients and pathogens.

This makes the contents similar in nature to sewage, and therefore potentially suitable to be treated in similar facilities as sewage. But there are significant differences we must take note of, as we will see in subsequent sections.

5.2 How sewage and septage/fecal sludge are different

In spite of the explanation given above, sewage and septage/FS are quite different. The origin of sewage and septage/FS is the same: excreta, mixed with ablutionary water/material, and possibly wash water from other domestic activities. Sewage, being water borne, is quickly and constantly conveyed to the treatment location. As such, its quality and quantity (and their variation) is quite well understood, and there is not much variation from context to context. The design basis for different sewage treatment facilities is therefore quite standard.

However, this is not so for FS or septage (together termed sludge in this guideline). There is wide variation in quality and quantity, depending on various factors. The most obvious differences are:

Sludge has higher (typically by an order of magnitude or more):

(a)   Solids (dissolved and suspended)
(b)   Biochemical oxygen demand (BOD)
(c)   Chemical oxygen demand (COD)
(d)   Nitrogen
(e)   Pathogens, particularly helminths
(f)   Fats, oils and grease

    (g)   Inorganic content (silt, sand and grit)

    (h)   Garbage/solid wastes

and sludge:

    (a)   is less easily biodegradable

    (b)   is highly variable (quantity and quality)

    (c)   may be potentially contaminated by toxic/industrial wastes

These factors cause variation in characteristics of sludge from containment to containment, depending on usage of toilets, locality to locality, from season to season, and also on frequency and method of emptying. Often emptying tankers also bring sludge from other sources, such as trade premises, commercial kitchens and restaurants, which further causes variations in sludge type.

Parameters that are typically considered for characterisation of sludge include solids concentration, BOD, COD, nutrients and pathogens. These parameters are the same as those considered for domestic sewage analysis. However, for sludge, the fractionalisation of the pollutants: particulate and dissolved, solids particle type and size profile, biodegradability, key ratios of parameters and presence of other inhibiting materials help show the differences in character between sludge and sewage.

5.3 Origin of sludge

Sludge is generated from the desludging or emptying of on-site containments. For properly designed and operated septic tanks, it is possible to calculate the amount of sludge that will accumulate over a period of time. This will depend on a number of factors including the number of people using the on-site system, its design and the frequency of emptying. Usually, however, the practice in most places is to empty the entire contents of a septic tank or other containment system. Then, the volume of sludge emptied is dependent only on the volume of the containment and the frequency of emptying.

Containments may be of different types:

(a) Properly designed septic tanks, with twin compartments, and supernatant overflow to soak-pits or filter or to surface drains. Usually such septic tanks serving a single household (of 5 people) would have a volume of 1.5–2.5 m^3. The solids settle at the bottom, and undergo anaerobic digestion, which reduces the quantity of solids over time. The supernatant exits the septic tank to the soak-pit or filter or to surface drainage. Over a two year period, up to 0.5 m^3 of sludge may accumulate in the tank. These containments should be desludged once in 2–3 years. Otherwise, the settled sludge will begin to overflow together with the supernatant into the soak-pit or filter, causing clogging, or to the surface drainage, causing pollution. Moreover, the accumulated sludge will reduce the effective volume of the containment, thereby reducing retention time and settling efficiency. Scheduled emptying may be appropriate for such septic tanks. The emptied material is dilute (<2% solids), and tends to be well digested. Where groundwater level is high, it may backflow into the septic tank, and the septic tank may need more frequent emptying, and the sludge tends to be even more dilute.

(b) Pits which are porous, without a base and where most of the liquid seeps away leaving a much more concentrated sludge. In conditions where groundwater level is low and soils are porous, the solids will settle and accumulate over long periods, and such containments may not require emptying for long periods, often 7–10 years. The period up to the first emptying will be prolonged, but after that, the soil pores get clogged by fine solids, and microbial growth takes place in the soil around the containment. Seepage is impeded, and the pit may fill faster. The emptied material may have high solids content (about 2–5%), and tends to be well digested. In conditions of high groundwater level, with backflows into the pit, it tends to fill up faster, and the emptied material tends to be more dilute.

(c)   Containments without overflow/outlet, which are emptied very frequently (often weeks to a few months). These are merely holding tanks, and little biological stabilisation occurs. The emptied material tends to be fresh, with high BOD and may be dilute.

(d)   Containments serving community toilets or public toilets or temporary/mobile toilet facilities tend to fill up very quickly and need to be emptied frequently, often every few days. Again, these are merely holding tanks, and little biological stabilisation occurs. The emptied material tends to be fresh, with high BOD and quite dilute.

Whatever the type of containment, these on-site systems merely remove part of the suspended matter and organics. The dissolved organics and most of the pathogens are not removed by the containments.

BOX 1 SUMMARY – ORIGIN OF SLUDGE

The points to note from this section are that:

- The quality and quantity of sludge is largely dependent on:
  - the type of containment and its usage/context
  - emptying frequencies and methods
- Interventions in fecal sludge management such as improved containments, scheduled emptying and stringent regulation will impact quality and quantity of sludge over time.

5.4 Solids in sludge

Sludge is mainly water, and this water may be in free or 'bound' forms. Free water is easily separated, while bound water is much more difficult to remove. Free water usually represents the bulk of water in untreated sludge. It can be separated from the solid phase by technologies such as settling or filtration. Water which is bound to solids is much more difficult to remove than free water and may need addition of chemicals or the use of centrifugation, pressure or evaporation to separate.

Sludge generally has very much higher solids content than sewage, often ranging from about 2,000 to over 50,000 mg/l. Solids content of septage tends to be low, while pit latrine sludge would be higher. Fresh sludges from public toilets will have values in the higher range.

The solids concentrations just mentioned are expressed as milligrams of suspended solids (SS) in one litre of sludge. In contrast, solids loadings are often more relevant, and this is obtained by multiplying volume by concentration. Depending on the context, the solids loadings should be considered in units of daily, hourly or annual loadings in kg/day, kg/hr or kg/year.

Total solids (TS) concentration of sludge is comprised of dissolved solids SS. TS consists of floating material (including garbage), settleable matter (including grit and sand), fine matter in suspension, colloidal material and matter in solution. Parameters that can be measured include TS, fractions of volatile or fixed solids, and settleable, suspended or dissolved solids as well as particle size distribution.

Some SSs are settleable, which means they can be separated by physical settling processes. Well-digested sludges (from pits and septic tanks) have higher settleability, while fresh sludges (from community toilets and container-type vaults) settle poorly. Solids that settle out of suspension after a certain period of time, for example, the solids that accumulate in the bottom of an Imhoff cone after 30–60 minutes, are termed settleable solids. This value is reported as the sludge volume index (SVI), and is used to help design settling tanks.

Sludge has different dewatering characteristics compared to STP sludge. The duration of on-site storage, and the age of the sludge affects the ability to dewater the sludge. It also contains large quantities of fats, oils and grease which are difficult to settle. 'Fresh' or 'raw' sludge is more difficult to dewater than older, more stabilised sludge.

The solids in sludge can be categorised as biodegradable or non-biodegradable. Pit sludges have the highest proportion of non-biodegradable or very slowly biodegradable portions, because they have been well digested in the years of residence in the pits. Similarly, septage from septic tanks also has a high proportion of non-biodegradable or slowly biodegradable portions, having undergone stabilisation in the septic tanks over the years. Fresh sludges have high biodegradability, and therefore, biological processes work well to stabilise them. Non-biodegradable solids are generally unaffected by biological processes. The ratio of volatile solids to total solids is used as an indicator of the relative amount of organic matter and the biodegradability of the sludge.

The particle size distribution, particle density, the type of particles (particulate, colloidal, floc, viscosity) all affect the settleability, dewaterability and oxygen transfer efficiency (from gas to liquid phase) and oxygen uptake (to the particulate organic matter). Therefore, the nature of the solids has a bearing on the treatability of the sludge.

Large discrete particles and flocs settle better than smaller or colloidal particles. This makes solids/liquid separation and dewatering easier for digested sludges. Colloidal particles that are not removed through gravity settling tend to be negatively charged, making them stable in suspension. Polymers are added that destabilise particles, allowing them to come in contact with each other, form larger flocs and settle, thereby achieving enhanced sedimentation.

In aerobic treatment reactors, oxygen is transferred from the air to the liquid, and high solids content accumulating in the reactors

Table 2 Characteristics of sewage and sludges.

	Sewage	**Septage**	**Public Toilet Sludge**
Characteristic	Medium strength sewage	Desludged after few years storage	High concentration, fresh, emptied after days of storage
TS	<1%	<3%	>3.5%
SS (mg/l)	200–700	7,000	>30,000

SS: suspended solids. *Source*: Data from Heinss, U., *et al.* (1998). Solids Separation and Pond Systems for the Treatment of Faecal Sludges in the Tropics. Lessons learnt and recommendations for preliminary design. SANDEC Report No. 5/98. Second Edition. Swiss Federal Institute for Environmental Science and Technology (EAWAG) and Water Research Institute (WRI), Accra/Ghana.

impedes oxygen transfer efficiencies. When particle sizes of solids are large or colloidal, oxygen uptake is also slower, making the stabilisation process slower.

Table 2 shows TS and SS concentrations for fresh sludge, septage and sewage. Actual values in the field may vary very widely.

BOX 2 SUMMARY – SOLIDS IN SLUDGE

The points to note from this section are that:

- solids in sludge are present at very much higher in concentration in sewage (20–100 times more)
- the difference between concentration of solids and total loadings of solids
- sludge is mainly water, and water may be in free or bound forms which need different processes for separation
- solids in sludge are present as dissolved solids or suspended solids
- separation of solids from the water may be done by different methods, which have varying separation efficiencies
- fresh sludges are more difficult to settle than well-digested sludges

- solids in sludge can be categorised as biodegradable or non-biodegradable
- fresh sludges contain higher portion of biodegradable solids than well-digested sludges
- particle size distribution, particle density, type of particles all affect the settleability, dewaterability and oxygen transfer efficiency and oxygen uptake

Interventions in fecal sludge management such as improved containments, scheduled emptying and stringent regulation will impact solids content and quality over time.

5.5 Organic matter in sludge

The organic content of sewage is usually measured using BOD. However, for sludge, COD, which is a more complete measure of the total organics present, is often used. COD measurements are recommended to be used since total COD can be subdivided into useful indicative fractions which have a bearing on the biological treatment processes. It should be noted that the organic matter may be in the form of SS or dissolved solids (see above).

Fresh sludge contains a high proportion of biodegradable material, part of it 'readily biodegradable' and the remainder 'slowly biodegradable'. On the other hand, digested sludge from pits and septic tanks has remained in the containments for many years, and is mostly stabilised, with the readily biodegradable portion already digested. It therefore contains a much higher proportion of non-biodegradable material. This is mostly particulate and hence potentially settleable.

Figure 3 shows a typical fractionalisation of COD for fresh and digested FS.

Sludge removed from frequently emptied public toilets and container-based sanitation systems will therefore be very dilute, have high portion of biodegradable COD and may have strong odour, besides having poor settleability. It may be appropriate to introduce bio-digestion as an initial step to stabilise the COD levels of such sludges before further biological treatment.

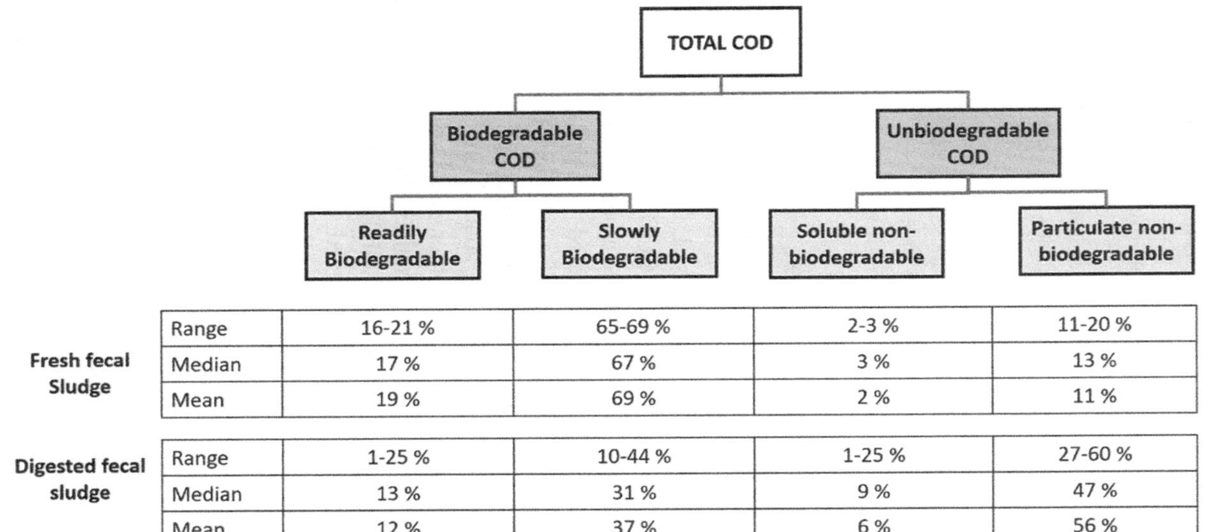

		Readily Biodegradable	Slowly Biodegradable	Soluble non-biodegradable	Particulate non-biodegradable
Fresh fecal Sludge	Range	16-21 %	65-69 %	2-3 %	11-20 %
	Median	17 %	67 %	3 %	13 %
	Mean	19 %	69 %	2 %	11 %
Digested fecal sludge	Range	1-25 %	10-44 %	1-25 %	27-60 %
	Median	13 %	31 %	9 %	47 %
	Mean	12 %	37 %	6 %	56 %

Figure 3 Chemical oxygen demand (COD) fractionalisation. *Source*: data from Lopez-Vazquez C., *et al.* (2014). Co-treatment of faecal sludge in municipal wastewater treatment plants, in Strande, L., *et al.* (eds.), *Faecal Sludge Management: Systems Approach for Implementation and Operation*, IWA Publishing, London. Table 9.3.

Septage from septic tanks, wet leach pits and wet pit latrines are also likely to have low biodegradability. The COD will be largely non-biodegradable, and comprise mainly particulate matter (which may also include a high proportion of biodegradable COD). Solid–liquid separation steps up front will help reduce much of the COD.

Sludge from pits too will be well digested, with low water content and low proportion of biodegradable COD, and a direct dewatering step may be applied.

The soluble non-biodegradable component of the COD will be mostly unaffected by the treatment process, and will pass through the process, and this could be a limiting value.

BOX 3 SUMMARY – ORGANIC MATTER IN SLUDGE

The points to note from this section are that:

- organic content of sludge is much higher than for sewage.
- COD can be subdivided into useful indicative fractions.
- fresh sludge contains a high proportion of biodegradable material; bio-digestion as an initial step may be appropriate
- digested sludge from pits and septic tanks contains more non-biodegradable material. However, this is mostly particulate and potentially settleable.
  - for dilute sludge, solid liquid separation step up front will help reduce much of the COD. Supernatant to be co-treated in STP.
  - for sludge from pits with low water content and low proportion of biodegradable COD, and a direct dewatering step may be applied. Liquid part to be co-treated in STP.
- The soluble non-biodegradable component of the COD will be unaffected by the treatment process and may be a limiting value.

Interventions in fecal sewage management such as improved containments, scheduled emptying and stringent regulation will impact organic content and quality of sludge over time.

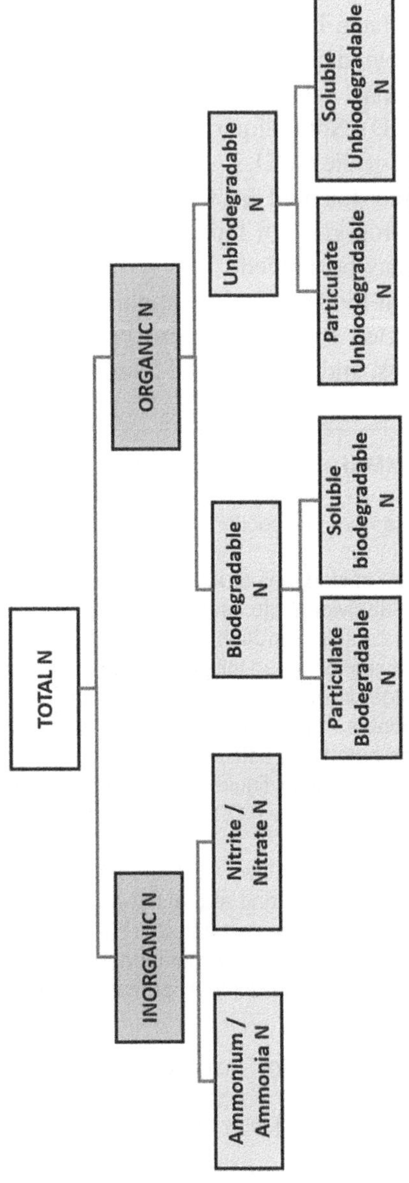

Figure 4 Nitrogen fractionalisation. *Source:* Data from Strande L., *et al.* (eds.) (2014). *Faecal Sludge Management: Systems Approach for Implementation and Operation.* IWA Publishing, London.

5.6 Nitrogen and phosphorus in sludge

Increasingly, effluent standards in many countries stipulate nutrient levels, and this would require reduction of the nitrogen and phosphorus levels. Sludge has much higher nitrogen concentrations than sewage, usually between 10 and 50 times more. Depending on the duration of storage and type of containments, sludge nitrogen could be in the form of ammonium (NH4-N), ammonia (NH3-N), nitrate (NO3-N), nitrite (NO2-N) or organic forms of nitrogen such as amino acids. Nitrogen removal is through the processes of nitrification (which requires oxygen) and denitrification (which happens in anoxic conditions but requires a carbon source). The organic content of the waste is the carbon source and a sufficient organic concentration is necessary for nitrogen removal by denitrification.

Non-biodegradable nitrogen cannot be removed through biological processes. However, the particulate non-biodegradable portion may settle out. The soluble non-biodegradable portion is unlikely to be affected by the treatment process, and will pass through the process, and this will be a limiting value.

Concentration of phosphorus is also much higher in sludge than in sewage, often by 2–30 times. Phosphorus is present as phosphate, the acid or base form of orthophosphoric acid (H_3PO_4/PO_4-P), or as organically bound phosphate. Phosphorus can be removed through precipitation, sedimentation, mineralisation or plant uptake in planted drying beds. Usually biological phosphorus removal will not be an issue in conventional treatment processes.

BOX 4 SUMMARY – NITROGEN AND PHOSPHORUS IN SLUDGE

The points to note from this section are that:

- Where effluent standards stipulate nutrient levels, nitrogen and phosphorus removal would be required.

- sludge has much higher nitrogen concentrations than sewage
- sludge nitrogen form varies depending on containment type and duration of storage.
- nitrogen removal through nitrification requires oxygen and denitrification requires a carbon source from the organic content. Therefore, sufficient organic concentration is necessary for nitrogen removal by denitrification.
- non-biodegradable soluble nitrogen cannot be removed through biological processes and will remain in the effluent. This will be a limiting value.

5.7 Pathogens

Sludge contains high levels of pathogens. This is, in part, because large cells, such as helminth eggs, settle and concentrate in the sludge. For sludge, helminths are commonly used as an indicator of the effectiveness of pathogen reduction.

Most STPs incorporate treatment technologies to separate solids and reduce the organic and SS loads in the liquid effluent. They will not produce an effluent suitable for reuse. Further treatment to remove pathogens will therefore be required if the liquid effluent is to be used for irrigation and will also be desirable if effluent is to be discharged to a water body that is used for recreation or as a source of potable water.

Effluent standards may stipulate coliform levels, and in cases where there is reuse of effluent and bio-solids, coliform and helminth levels may be stipulated.

Separated liquid will require further treatment to reduce pathogen numbers to safe levels, particularly where the treated effluent is to be reused. Among options are lagooning, chlorination, ozone and UV treatment.

Similarly, dewatered solids may require further treatment to remove pathogens to render biosolids suitable for reuse. Among options are storage for extended period, composting, lime stabilisation, infrared radiation and thermophilic biodigestion.

5.8 Volume of sludge

The quantity of sludge we are concerned with is the volume of sludge arriving at the treatment facility. The practice in most places is that septic tanks or other containments are emptied, and not desludged. In other words, the entire contents of the tank are removed. This being the case, the quantity of sludge is dependent only on the frequency of emptying and the volume of the containment tank.

In general, the volume of sludge to be handled in a year can be estimated from the total volume of containments, divided by the average frequency (in years) of emptying. The sludge will be transported by tankers to the treatment facility, and the volume of tankers and the frequency of arrival at the facility will determine the hourly volume of sludge to be handled. The volume will vary based on seasonal variations in emptying, tanker size and operating hours of the tankers and treatment facility.

The total volume of sludge produced by a community varies greatly depending on the type of containments, groundwater infiltration and emptying frequency.

The realistic situation we can expect in an urban area would be a combination of the different types of containments, which would result in different volumes of fresh and stabilised sludge, partly from septic tanks, partly from pits, container vaults and community toilets. The composition and relative volumes should be taken into account when assessing solutions.

BOX 5 VOLUME OF SLUDGE AND SEWAGE FROM DIFFERENT CONTAINMENT TYPES

Table below is indicative of several common scenarios for a community of 1,000 households:

Scenario	Containment Volume	Emptying Frequency	Volume of Sludge Per Annum
Septic tanks, with overflow to soak pit, low ground water, permeable soils	2 m^3	3 years	$(1{,}000/3) \times 2 = 667$ m^3
	3 m^3	5 years	$(1{,}000/5) \times 3 = 600$ m^3
Pits, with liquid seeping to soils, low ground water, permeable soils	2 m^3	10 years	$(1{,}000/10) \times 2 = 200$ m^3
Pits with high ground water	2 m^3	1 year	$(1{,}000/1) \times 2 = 2{,}000$ m^3
Containment vaults with no outlet	10 m^3	3 months	$(1{,}000/(3/12)) \times 10$ $= 40{,}000$ m^3
Community toilets (5 containments)	150 m^3	1 month	$(5/(1/12)) \times 150$ $= 9{,}000$ m^3
Sewered system (sewage, with grey water inclusion)			$1{,}000 \times 5 \times 135 \times 0.8 \times 365$ $= 197{,}100$ m^3

Assumptions:

(1) No of households: 1,000
(2) Average household size: 5
(3) Average containment size as in column 2
(4) Average frequency of emptying as in column 3
(5) Water consumption: 135 litres/capita/day
(6) Sewage generation: 80% of water consumption (with grey water inclusion)

The volume of sludge varies for different types of containments, depending on the design, size and frequency of emptying. Also, the equivalent number of households using a sewer system generates a much larger volume of sewage (assuming grey water is also included).

Another aspect to bear in mind is the variation of volume of sludge arriving at the treatment facility over the day based on tanker arrival and discharge rates, usually confined to working days/hours. In actual practice, the sludge will be discharged based on tanker volume, and within a time frame of about

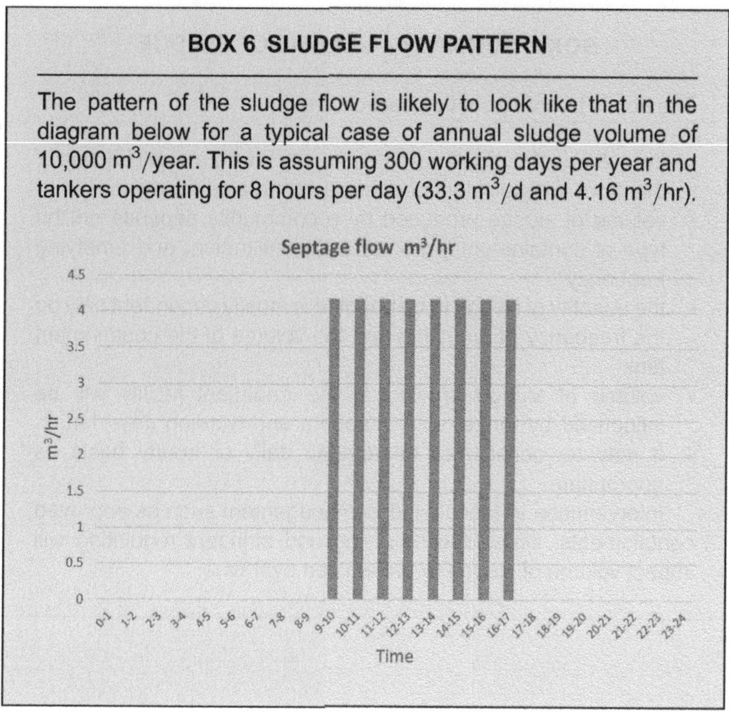

BOX 6 SLUDGE FLOW PATTERN

The pattern of the sludge flow is likely to look like that in the diagram below for a typical case of annual sludge volume of 10,000 m^3/year. This is assuming 300 working days per year and tankers operating for 8 hours per day (33.3 m^3/d and 4.16 m^3/hr).

10 minutes, creating sharp peaks in sludge arrival rates. A holding tank will be appropriate, to blend and equalise the flow.

For purposes of estimation of current quantities, and to estimate projected flows over the planning period, it may be necessary to also consider:

- current collection rate
- available collection, transport and treatment/disposal infrastructure
- logistics (distance from collection area to treatment facility)
- expected population growth
- expected changes in regulatory framework
- expected changes in containment designs
- costs, fees
- other factors.

BOX 7 SUMMARY – VOLUME OF SLUDGE

The points to note from this section are that:

- the volume of sludge from a community is only a small fraction of that from an equivalent community using a sewer system
- volume of sludge produced by a community depends on the type of containments, ground water infiltration, and emptying frequency
- the quantity of sludge to be handled is mostly dependent only on the frequency of emptying and the volume of the containment tank
- volume of sludge arriving at the treatment facility will be influenced by tanker size, numbers and wording days/hours. It may be considered on annual, daily or hourly basis as appropriate.

Interventions in faecal sludge management such as improved containments, scheduled emptying and stringent regulation will impact volume of sludge to be handled over time.

6 IDENTIFYING CO-TREATMENT OPPORTUNITIES

For cities where sewerage coverage is very limited and the majority of households rely on on-site sanitation, the ratio of sludge to sewage flows is likely to be relatively high, and a dedicated sludge treatment facility may be a better option.

Cities with an STP, where a substantial percentage of population is connected to the STP, but where the STP still has significant spare unutilised capacity (which is unlikely to be used up in the near future), are good candidates for co-treatment options. Actual local conditions should be considered, especially those relating to sewage flows and projections as well as sludge flows and projections. The additional sludge volume co-treated should never exceed the spare unutilised capacity of the STP.

When considering co-treatment, the co-treatment facility, its process units and design limitations (such as hydraulics, solids loading, organic loading, oxygen requirement, etc) should be considered against the impact of the characteristics of the sludge that will be processed in it. The critical process units which create limiting conditions should be identified, and a decision made whether to co-treat. If it is decided to co-treat, the following options could be considered:

- Limit the quantity of sludge to be co-treated
- Pre-treat the sludge to modify its characteristics, to mitigate the limiting conditions in the co-treatment facility
- Retrofit the limiting process units in the treatment facility to enable the treatment facility to handle the sludge
- Combination of the above

The subsequent sections of this guide elaborate on these.

The time dimension should also be borne in mind – how the situation will change over time:

- Changes of sewage flow to the STP with population growth or increased connections

- Changes of sludge flow due to increased number of containments or more efficient or organised sludge collection
- Changes in characteristics of sludge : when the ratio of different types of containments changes, sludge characteristics will change

Often the most convenient co-treatment solution would appear to be to discharge the sludge at sewer manholes or pump stations of the sewerage system. While this method is attractive because of convenience and logistical advantages, there are serious consequences which should be considered:

- When crude sludge is directly emptied into manholes or pump stations, there will be deposition of solids because the sewers are designed for normal domestic sewage, with solids contents between 200–500 mg/l rather than the high solids content (including silt and garbage) of sludge, and blockage in sewers and pump stations and excessive wear and tear on pumps and abrasion of pipes will happen.
- High fats, oils and grease in the sludge will cause clogging and blockages in pipelines and equipment.
- Control and monitoring may be difficult if the emptying is done at remote locations. There are high risks of toxic or incompatible wastes being dumped into sewerage systems.
- The capacity of the sewers and pump stations should be adequate to handle the additional flow (considering the sporadic peaks expected in the discharge of the sludge).

For such discharge of sludge to the sewerage system, decanting stations with proper facilities for screening, grit removal and if possible, solids–liquid separation are recommended. Remote decanting stations should also have adequate monitoring facilities for the sludge being added.

Similarly, in cases where the discharge of sludge for co-treatment is made directly at the STP, proper facilities for screening, grit removal, fats/oil/grease removal, equalisation/blending and other preliminary treatment processes as required should be provided. Solids—liquid separation is an essential preliminary step in co-treatment and should be incorporated, with the liquid part co-treated with the sewage.

Exceptions to the above may be considered in the cases where the sludge is extremely dilute with little grit/garbage, and the volume is very small.

It should be noted that solids/liquid separation will remove a large portion of the particulate matter in sludge, and a corresponding portion of the organic matter. However, the liquid from solids/liquid separation process will still be very much higher in organic and SS than sewage, and requires further treatment.

Tayler (2018) suggests the following treatment steps for sludge treatment, and these should be borne in mind in parallel as we consider co-treatment:

- Removal of gross solids, grit, fats, oil and grease, and floating objects.
- Stabilisation of fresh FS to reduce odours and render it more amenable to follow-up treatment processes.
- Solids/liquid separation.
- Treatment of the liquid removed from septage or FS.
- Solids dewatering.
- Reduction of the pathogen content of treated liquid and separated sludge.

7 IMPACTS OF CO-TREATMENT ON SEWAGE TREATMENT FACILITIES

When sludge is added to the sewage stream, it may impact the STP in several ways:

- odour issues, especially at the sludge reception area (particularly for fresh sludges)
- increase in the quantity of screenings and grit
- increase in sludge and scum
- significantly higher organic and solids loadings
- higher nitrogen
- potential shock loading due to irregular addition of sludge
- potential toxic substances in sludge

The STP receives sewage at a fairly consistent rate (subject to daily variations and peaks), but sludge flows vary widely, depending on tanker volume, discharge frequency, working hours of tanker operations, seasonal variation of emptying, etc. All these do not really impact the hydraulics of the STP, if the relative volumes of sludge are very low. However, in terms of loadings: BOD, COD, solids and ammonia, the intermittent discharges can cause serious shock loadings and process upsets. The nature of sludge can vary from tanker to tanker.

Table 3 below shows the comparison of typical characteristics of fresh sludge, septage and sewage.

Table 3 Characteristics of fecal sludges and comparison with tropical sewage.

Characteristics	Sewage	Septage	Public Toilet Sludge
	Tropical Sewage	**Low Concentration, Well Stabilised**	**High Concentration, Mostly Fresh**
COD (mgl)	500–2,500	<10,000	20,000–50,000
COD/BOD	2:1	5:1–10:1	2:1–5:1
NH4–N (mgl)	30–70	<1,000	2,000–5,000
TS	<1%	<3%	>3.5%
SS mgl	200–700	7,000	>30,000
Helminth eggs (no/litre)	300–2,000	4,000	20,000–60,000

Source: Data from Heinss, U., *et al.* (1998). Solids Separation and Pond Systems for the Treatment of Faecal Sludges in the Tropics. Lessons learnt and recommendations for preliminary design. SANDEC Report No. 5/98. Second Edition. Swiss Federal Institute for Environmental Science and Technology (EAWAG) and Water Research Institute (WRI), Accra/Ghana.

8 PLANNING FOR CO-TREATMENT

When planning for co-treatment (Figure 5), the following information/data should be available for assessment of technical viability. In the absence of reliable data, appropriate conservative figures may be estimated or adopted from literature for planning purposes.

(a)   the characteristics of the STP where co-treatment is being proposed
  - size of STP: should be sufficient to mitigate shock loadings from tanker discharge volumes, or alternatively, justifiable to allow investment on reception facility, screening/grit removal, blending/mixing and possibly solids/liquid separation
  - spare capacity, expected to be available for planning period to accommodate the sludge to be co-treated
  - Key process design parameters, including sizing, retention times, surface overflow rates, oxygen supply, mixed liquor volatile suspended solids (MLVSS), food to microorganism ratio, sludge age, equipment ratings and STP sludge handling capacity
  - Regulated effluent standards (BOD, COD, SS, N, P, coliforms, others)
  - Current STP effluent performance (BOD, COD, SS, N, P, coliforms, others). The STP should be meeting the effluent standards.

(b)   Characteristics and future prospect of the catchment of the STP
  - Current sewage flows (daily, peak)
  - Projection of sewage flows over planning period
  - Characteristics of sewage and expected changes in planning period

(c)   Characteristics and future prospect of the sludge catchment for sludge to be co-treated at the STP

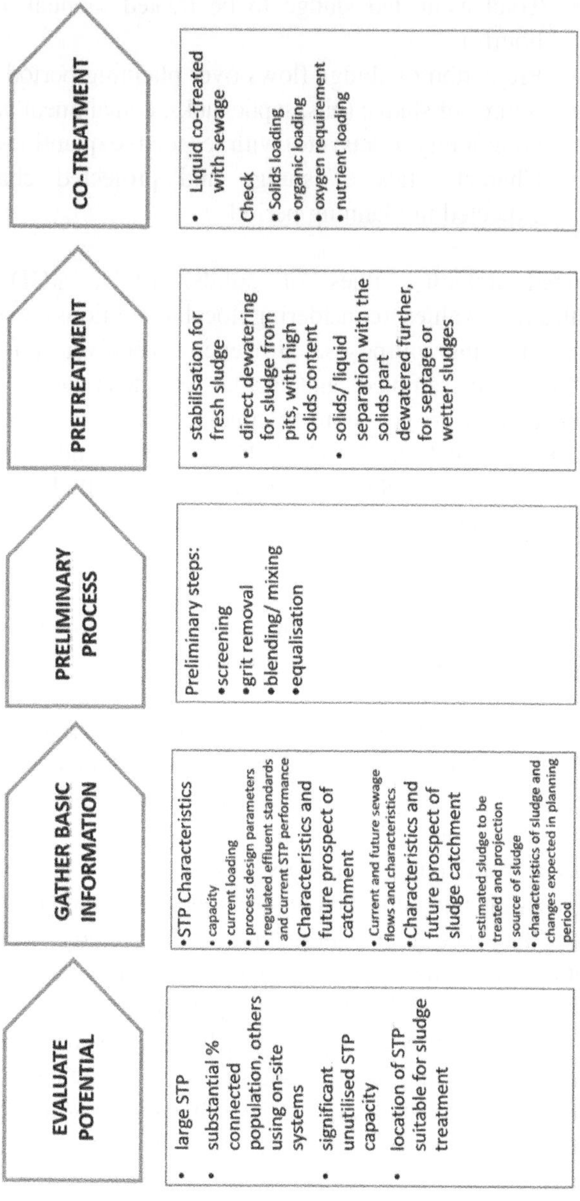

EVALUATE POTENTIAL

- large STP
- substantial % connected population, others using on-site systems
- significant unutilised STP capacity
- location of STP suitable for sludge treatment

GATHER BASIC INFORMATION

- STP Characteristics
  - capacity
  - current loading
  - process design parameters
  - regulated effluent standards and current STP performance
- Characteristics and future prospect of catchment
  - Current and future sewage flows and characteristics
- Characteristics and future prospect of sludge catchment
  - estimated sludge to be treated and projection
  - source of sludge
  - characteristics of sludge and changes expected in planning period

PRELIMINARY PROCESS

Preliminary steps:
- screening
- grit removal
- blending/ mixing
- equalisation

PRETREATMENT

- stabilisation for fresh sludge
- direct dewatering for sludge from pits, with high solids content
- solids/ liquid separation with the solids part dewatered further, for septage or wetter sludges

CO-TREATMENT

- Liquid co-treated with sewage

Check
- Solids loading
- organic loading
- oxygen requirement
- nutrient loading

Figure 5 Planning for co-treatment.

- Total estimated sludge to be treated (annual, daily, hourly)
- Projection of sludge flows over planning period
- Source of sludge (pits, septic tanks, containment vaults, community toilets, etc) with respective quantities
- Characteristics of sludge and projected changes expected in planning period

Estimated default values of solids, BOD, COD and fractionalisation values (considering local conditions) may be used for planning purposes. However, wherever possible, these should be sampled and actual characterisation done from time to time to get more reliable local values. Ratios of key parameters such as BOD/COD, TS/VSS, indicators of biodegradability, settleability and ammonia will be useful indicators to understand the nature of the sludge.

Preliminary steps such as screening, grit removal and blending/mixing, and where possible, solids/liquid separation, and balancing tanks to blend the sludge and equalise it, by mixing into the sewage to avoid shock loads, should be provided as appropriate.

This will enable a large portion of the garbage, grit, organic solids (and its BOD/COD) to be removed, bringing the FS characteristics closer to those of sewage, and therefore more manageable through co-treatment. Operational problems such as high garbage/grit/ solids accumulation in preliminary/primary stages of the STP (and even sewers and pump stations) can then be avoided.

Where the incoming sludge is predominantly fresh, a stabilisation step may be appropriate before solid liquid separation or as part of solids/liquid separation.

Where the sludge is from pits, and has high solids content, direct dewatering could be done and the liquid co-treated.

Septage or wetter sludges may be subjected to solids/liquid separation with the solids part dewatered further, and the liquid part co-treated.

Solid–liquid separation or dewatering will greatly reduce solids, and a large part of the particulate organics. From the solid liquid separation or dewatering step mass balance, the remaining organics in the liquid can be assessed.

STPs hosting co-treatment are likely to be mechanised plants utilising aerobic processes. Common systems would be activated-sludge based. The stages may include primary solids settlement, followed by an aerobic reactor to biologically stabilise organics, including possibly staged removal of nutrients. Final clarification processes will remove most of the solids, producing effluent with low levels of solids and organics. A disinfection step may be incorporated to remove pathogens, and other polishing steps may be included too.

Other common sewage treatment systems are pond systems, SBRs (sequential batch reactors), which is a variation of the activated sludge system, but operating on a batch basis, and UASBs (upflow anaerobic sludge blanket reactors).

The main part of the following guide refers to co-treatment in an activated sludge STP. Co-treatment considerations for other systems such as ponds, SBRs and UASBs are briefly discussed subsequently.

9 PRELIMINARY PROCESSES

Besides screening, grit removal and blending/balancing, some of the other preliminary processes that may be required before the co-treatment of the sludge are stabilisation and solids/liquid separation, as described in the next two sections.

9.1 Stabilisation

Stabilisation is often required for sludge sourced from simple containments, such as public or community toilet containments, which are emptied at very frequent intervals and where the sludge is fresh. Typical characteristics of fresh sludge are shown in Table 4.

This type of sludge needs to be stabilised before subsequent co-treatment. Available methods of stabilisation of sludge are lime stabilisation, aerobic digestion and anaerobic digestion. Tayler (2018) recommends anaerobic digestion, considering low energy requirements and also effectiveness. Small-scale biodigesters are suggested, due to their simplicity and because they do not need a power supply. These are most effective for sludges with high solids content and biodegradable solids. The design should facilitate mixing of the contents. This step will stabilise the bulk of biodegradable material, while also helping to blend and homogenise the sludge. Digested sludge will also

Table 4 Characteristics of fresh sludge.

Parameter	Range
COD	>20,000 mg/l
COD:BOD ratio	less than 5:1
Ammonia	>2,000 mg/l
SS	>15,000 mg/l

have better settleability and dewaterability. The BOD reduction varies according to the input sludge characteristics and the digester design. In the absence of data, for predominantly fresh sludge, digestion may be assumed to reduce at least 30–40% of BOD. Further details and design considerations of digestion are available in Tayler (2018).

9.2 Solids/liquid separation

Stabilised sludge sourced predominantly from septic tanks, pits or other containments will have good settling/dewatering characteristics. Such sludges are likely to have large proportion of non-biodegradable COD, which is mostly particulate and settleable. This makes solid/liquid separation a desirable first step for such sludges. Depending on the type of process adopted, a large portion of the solids and the BOD will be removed in this step.

Table 5 shows indicative reduction of solids and BOD in the solids/liquid separation stage.

The supernatant will need to be biologically stabilised in the aeration reactor with the sewage, and the solids settled out. The thickened sludge may be dewatered and further treated before disposal/reuse. Tayler (2018) has more information on solids/liquid separation and solids dewatering.

Table 5 Indicative reduction of solids and BOD in the solids/liquid separation stage. *Source*: Data from Tayler (2018).

	% TSS Reduction	% BOD Reduction
Drying beds	95	70–90
Anaerobic ponds	80	60
Belt press	95	90
Gravity thickening	30–60	30–50

BOX 8 MASS BALANCE

The diagram below shows a typical mass balance diagram, assuming sludge volume of 18,250 m^3/year (60.83 m^3/day based on 300 days operation per year), sludge SS at 10,000 mg/l, gravity thickening with 50% TSS reduction to 5% solids content. Note that 50% solids removal refers to the solids loading in kilograms and not to the concentration in mg/l.

See Excel sheet 1 for example of mass flow for solids/liquid separation.

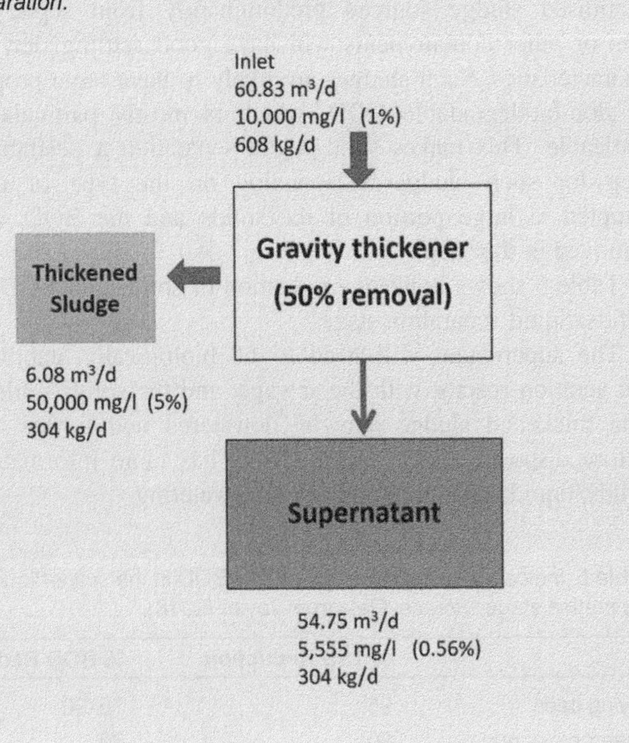

9.3 Solids/liquid separation in primary settling tank

In the cases where the STP incorporates primary settlement, the solids/liquid separation may be performed in the same tank. This should be done with caution, especially for old facilities. Hydraulic impact during peak flows, and additional solids loading impact should be checked.

Sewage flows vary over the day, with peak flows in the morning and evening. Peak factors are generally lower for larger STPs, since the flow travels for a longer period along sewer pipelines before reaching the treatment facility, and is therefore more balanced.

If no equalisation is provided, the morning peak will generally coincide with the hours when sludge is delivered to the facility, when the flow would be the combination of the hourly sewage peak with the sludge flow. The flow patterns of the sewage, sludge and combined flow are likely to look as shown in Figure 6.

As we can see, the critical period is when the morning peak sewage flow coincides with the sludge flow. Surface overflow rate in the primary clarifier should be checked on the hourly

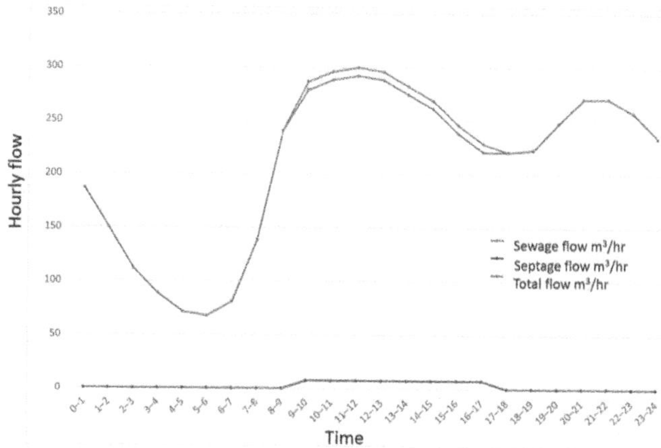

Figure 6 Hourly flow pattern with co-treatment.

peak flow of the combined sewage and sludge flow. Usually at low percentage of sludge addition this should not impact greatly. Moreover, septage and pit sludge usually have good settling properties. However, if there is large percentage of fresh sludge, settleability could be a problem.

Recommended good practice in co-treatment is to ensure that the host STP has sufficient spare capacity, and this ensures the design surface overflow rates are unlikely to be exceeded at low percentages of sludge co-treatment.

BOX 9 HOURLY HYDRAULIC LOAD IN PRIMARY SETTLING TANK

Combined hydraulic load $Q_t = Q_w + Q_s$

where:

Q_t = total flow;
Q_w = peak sewage flow;
Q_s = septage flow;

Assuming an STP of $10,000\,\mathrm{m^3/day}$ design flow, which is currently 50% utilised:

Current sewage flow $Q_w = 10,000 \times 50\% \times 365$

$$= 1,825,000\,\mathrm{m^3/year}$$

If 1% of this flow is added as sludge:

$Q_s = 1,825,000 \times 0.01 = 18,250\,\mathrm{m^3/year}$

Hourly sludge volume is calculated considering 300 working days per year and 8 working hours per day for sludge tanker reception:

Hourly sludge flow: $Q_s = \dfrac{18,250}{300 \times 8} = 7.6\,\mathrm{m^3/hr}$

Peak sewage flow is calculated using the modified Babbitt formula:

Peak flow factor $= 4.7 \times \left(\dfrac{p}{1,000}\right)^{-0.11}$

where P is the population equivalent, in this case: $(10{,}000/135 \times 0.8) = 92{,}600$ people

$$\text{Peak flow factor} = 4.7 \times \left(\frac{92{,}600}{1{,}000}\right)^{-0.11} = 2.86$$

Design peak flow $= 2.86 \times 10{,}000/24 = 1190\,\mathrm{m^3/hr}$

Current peak flow $= 2.86 \times 5{,}000/24 = 595\,\mathrm{m^3/hr}$

(Peak flow factor is calculated on total design population, but current peak flow is based on 50% loading as assumed.)

Total combined peak flow for co-treatment condition

$$= 595 + 7.6 = 603\,\mathrm{m^3/hr}$$

The increase is $(603 - 1{,}190)/1190 = 49\%$.

Note: The percentage is over the design peak flow. Hydraulic overloading is usually not an issue at low percentages of sludge addition, and where there is sufficient spare unutilised STP capacity.

See Excel sheet 2.

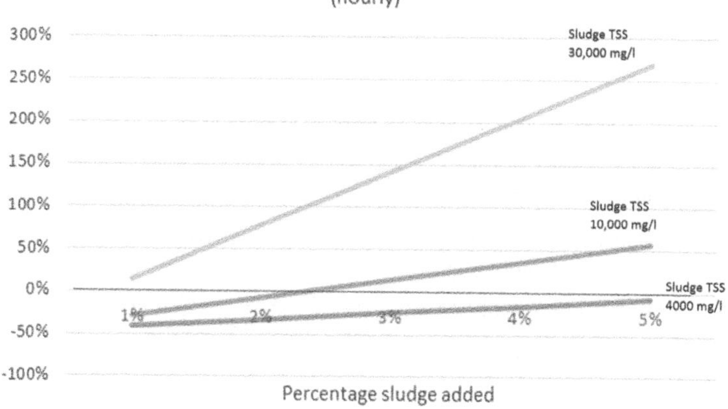

Figure 7 Percentage increase in primary sludge (hourly).

There will also be more sludge produced, which will need to be pumped and handled.

The graph in Figure 7 shows the percentage increase in hourly primary sludge for various incoming sludge TSS and percentage sludge addition. This is for an STP with 50% unutilised capacity.

The increase of solids removed will be very high for higher sludge concentrations and for higher percentage sludge added. STPs with lower unutilised capacity will also be seriously impacted. Sludge handling equipment and primary sludge treatment facilities will need to checked and upgraded/expanded accordingly.

BOX 10 INCREASE IN PRIMARY SETTLING TANK SLUDGE

According to Metcalf & Eddy (1991), primary settling tanks can remove 50–70% of suspended solids. Assuming a 50% solids removal, at peak flow conditions:

Solids removed under design conditions: $= 50\%$ of $(Q_w C_w)$

where:

Q_w = design peak hourly sewage flow (1190 m^3/hr, see Box 9);
C_w = concentration of TSS in the sewage (300 mg/l);

$$\text{Solids removed} = 50\% \text{ of } (1{,}190 \times 300)/1{,}000 \text{ kg}$$
$$= 178.5 \text{ kg/hr}$$

Solids removed under co-treatment conditions:

$$= 50\% \text{ of } (Q_w C_w + Q_s C_s)$$

where:

Q_w = current peak hourly sewage flow (595 m^3/hr, see Box 9);
Q_s = sludge flow (7.6 m^3/hr);
C_w = concentration of TSS in the sewage (300 mg/l);
C_s = TSS concentration in the liquid fraction of sludge (10,000 mg/l)

$$\text{Solids removed} = 50\% \text{ of } (595 \times 300 + 7.6$$
$$\times 10{,}000)/1{,}000 \, \text{kg}$$
$$= 127.3 \, \text{kg/hr}$$

This is less than the design solids by 29%. For higher sludge concentrations and for higher percentage sludge added, the increase of solids removed will be drastically higher. For STPs with lower unutilised capacity, the increase of solids will be greater as well. Sludge handling equipment and primary sludge treatment facilities will need to checked and be upgraded/ expanded accordingly.

See Excel sheet 2.

10 CO-TREATMENT OF LIQUID STREAM

The supernatant from solids liquid separation will need to be biologically stabilised in the aeration reactor. Likely impacts on liquid stream processes are:

(1) Insufficient aeration capacity for combined sewage/sludge load
(2) Lower oxygen transfer rate and uptake due to higher solids in reactor (both dissolved and suspended)
(3) Lower mixing effect due to high solids content
(4) Higher solids loading and residence time in reactor
(5) Higher surface overflow rate of clarifier, considering possible lower settleability
(6) Higher oxygen uptake for nitrification
(7) Unavailability of sufficient organic carbon for denitrification

10.1 Increase of flow volume and variation over the day

Combined hydraulic load is given by:

$$Q_t = Q_w + Q_s$$

where:

Q_t = total daily flow;
Q_w = daily sewage flow;
Q_s = daily septage flow;

We saw in previous sections that even when equalisation is not provided for the sewage flows and sludge flows, the increase of peak flows for the combined flow is unlikely to be significant. This is especially so considering that the STP where co-treatment is planned is expected to have significant spare capacity.

In processes with short retention times, instantaneous flows or at least hourly peak flows should be checked. For complete mix biological reactors with long retention times, daily average flows are more relevant and hourly variations are less significant.

The reduced retention time in the aeration reactor due to the marginally increased hydraulic load will be insignificant for reasonably low percentages of sludge co-treated (the volume of sludge added will be much less than the total spare capacity available in the STP).

10.2 Increase of solids loading and variation over the day

The solids loading in the combined flow is considered after reduction due to solids/liquid separation (or primary settlement). On a daily loading basis, the percentage increase of solids loading is given by:

Percentage increase of daily solids loading

$$= \frac{\text{Total increase in solids per day under co-treatment condition}}{\text{Total solids per day under design condition}}$$

$$\times\ 100\%$$

Considering sewage flow variation over the day (and corresponding SS variation) and sludge flow variation based on tanker arrival and discharge rates (in the event there is no equalisation provided), the hourly solids loading patterns of the sewage solids, sludge solids and combined solids are likely to appear as shown in Figure 8 (the dark line shows the combined flow, calculated as a moving average to allow for any attenuation due to reception facility, storage etc). It can be seen that in the absence of equalisation, there is a shock increase of solids during the period of sludge discharge. Process control should be carefully managed to adjust sludge return and wasting rates and reactor mixed liquor suspended solids (MLSS) to maintain stable conditions during this period.

However, in most completely mixed systems, with processes having long retention periods (such as activated sludge and

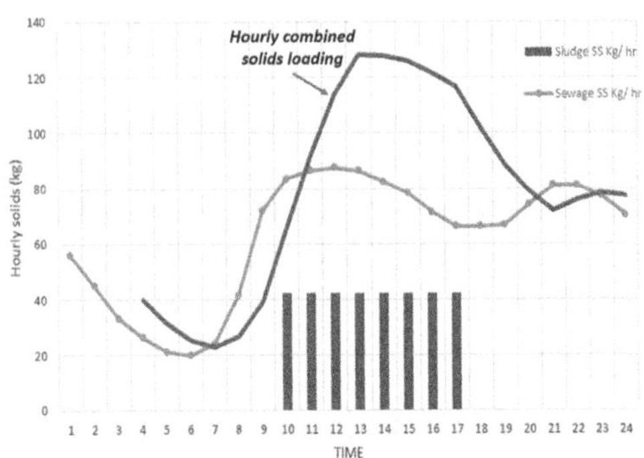

Figure 8 Solids loading under co-treatment conditions. SS: suspended solids. *Note: Hourly combined solids loading, calculated as moving average to allow for any attenuation due to reception facility, storage, etc.*

extended aeration systems), the impact is damped and not so pronounced. It is therefore sufficient to consider increase in solids loading on a daily basis for process design check.

BOX 11 INCREASE OF SOLIDS LOADING (DAILY)

Percentage increase of daily solids loading

$$= \frac{(Q_w \times C_w + Q_s \times C_s) - (Q_{wd} \times C_w)}{(Q_{wd} \times C_w)}$$

where:

Q_{wd} = design daily sewage flow 10,000 m^3/d;
Q_w = current daily sewage flow 5000 m^3/d;
Q_s = daily sludge flow 54.75 m^3/d;
C_w = sewage solids concentration (entering reactor) 300 mg/l;
C_s = sludge solids concentration (entering reactor, after solids/ liquid separation) 5556 mg/l

$$= \frac{(5,000 \times 300 + 54.75 \times 5,556) - (10,000 \times 300)}{(10,000 \times 300)}$$

$$= -40\%$$

> In this case, the solids loading under co-treatment condition is less than the design condition.
>
> However, in cases where percentages of sludge added, sludge TSS and STP current utilisation is high, the solids loading can be significantly large.
>
> *See Excel sheet 3.*

The graph in Figure 9 shows the estimated daily increase in incoming solids for various incoming sludge TSS and percentage sludge addition.

The impact will be in the following aspects:

(a)   increased load on aerators for mixing of aeration tank
(b)   reduced oxygen transfer rate
(c)   increased pumping for return sludge and excess sludge
(d)   increased volume of secondary sludge
(e)   impact on secondary settling tanks: problems with solids/ liquid separation, solids being washed out in the effluent

Based on Figure 9, the impact is high for high percentage of sludge being co-treated, especially where the sludge TSS is high. With solids/liquid separation, the sludge TSS can be

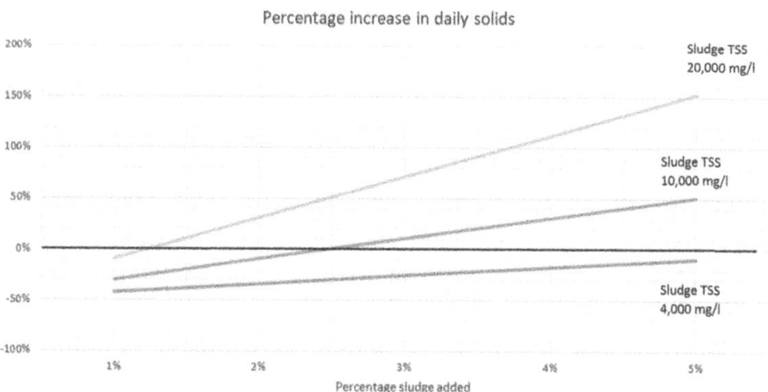

Figure 9 Percentage increase in daily solids loading.

reduced and the impact mitigated. Otherwise the aerator/mixer and sludge pump capacity should be reviewed and upgraded if necessary. The secondary sludge facilities may also need to be upsized.

10.3 Increase of organic loading and nutrients

Depending on the method of solids/liquid separation, a portion of the COD/BOD will also be reduced, with the particulate non-biodegradable and slowly biodegradable portions largely removed. The soluble portions and the readily biodegradable portions (which are likely to be fine particles not easily settleable/separable) are likely to remain in the liquid stream.

An estimate of the remaining COD and its fractions can be made, and the biological reactor designed to treat that. At the same time, ammonia compounds will also be nitrified in most tropical climates, and oxygen will be required for that too. As such the total oxygen required should consider the total BOD to be removed and total Kjeldahl nitrogen (TKN) to be nitrified.

Daily flow and loadings may be used for purpose of design of reactors, especially since these are likely to have long retention periods. To account for lower oxygen transfer and uptake due to higher solids concentration in the reactor, a correction factor is proposed. A correction factor sliding between 0.70 and 0.95 is applied in consideration of reduced oxygen transfer rates, depending on degree of solids concentration. Metcalf & Eddy (1991) recommends 0.7–0.95 range to correct for normal sewage, with 0.95 recommended for sewage. We recommend the lower end of range, 0.7 for co-treatment situations with at least 1% sludge added, sliding to 0.95 for 0% sludge added.

This oxygen requirement under co-treatment condition is compared with the design oxygen supply for the STP to check adequacy (Table 6). If the actual aeration capacity of the STP is unknown, as an approximation, the oxygen required to remove the BOD and TKN of the combined sewage and sludge can be compared with the oxygen required based on removal of design sewage BOD and TKN.

The graphs in Figure 10 were developed based on the equation in Table 6, indicating maximum volumetric sludge that may be co-treated, as a percentage of current sewage flow, for STP

Table 6 Oxygen requirement.

	Design Condition	Co-treatment Condition
Total daily BOD load to be removed	$B_r = Q_{wd} (B_w - 50)$ mg/l	$B_r = Q_{wu} (B_w - 50) + Q_s (B_s - 50)$ mg/l
Total daily TKN load to be removed	$N_r = Q_{wd} (N_w - 10)$ mg/l	$N_r = Q_{wu} (N_w - 10) + Q_s (N_s - 10)$ mg/l
Oxygen required $B_r \times 1.5 + N_r \times 4.57$	$1.5 \times Q_{wd} (B_w\ 50) + 4.57 \times Q_{wd} (N_w - 10)$	$1.5 \times (Q_{wu} (B_w - 50) + Q_s (B_s - 50)) + 4.57 \times (Q_{wu} (N_w - 10) + Q_s (N_s - 10))$
		The oxygen required will be adjusted, dividing by the correction factor of between 0.7 and 0.95

where:
B_r = BOD concentration to be removed;
Q_{wd} = STP design sewage flow;
Q_{wu} = current STP sewage flow;
Q_s = septage flow;
B_w = concentration of BOD in the sewage;
B_s = BOD concentration in the liquid fraction of separated septage/FS;
N_r = NH$_4$ concentration to be removed;
N_w = concentration of NH4 in the sewage; and
N_s = NH$_4$ concentration in the liquid fraction of separated septage/FS
Effluent standard for BOD is assumed at 50 mg/l, and TKN at 10 mg/l
Oxygen required for BOD removal: 1.5 kg O$_2$/kg BOD removed
Oxygen required for TKN nitrification: 4.57 kg O$_2$/kg TKN nitrified

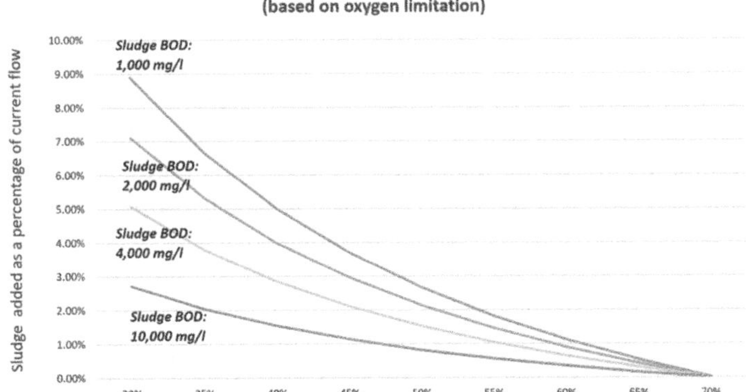

Figure 10 Maximum co-treatment percentage based on oxygen limitation (constant correction factor of 0.7 used). BOD: biochemical oxygen demand.

utilisation rates from 30% to 70%, and for incoming BOD concentrations from 1,000 to 10,000 mg/l. A constant correction factor of 0.7 was used to adjust the oxygen required for this graph.

We can see that oxygen limitations will restrict the percentage of sludge to be co-treated, especially where the current STP utilisation is high and the BOD of sludge is high. These

---

BOX 12 OXYGEN REQUIREMENT

Design condition : $1.5 \times Q_{wd}(B_w - 50) + 4.57 \times Q_{wd}(N_w - 10)$

Co-treatment condition (oxygen required will be adjusted, dividing by the correction factor of 0.7 to 0.95)

$$\frac{1.5 \times (Q_{wu}(B_w - 50) + Q_s(B_s - 50)) + 4.57 \times (Q_{wu}(N_w - 10) + Q_s(N_s - 0))}{Cf}$$

where:

Q_{wd} = STP design flow (10,000 m^3/d);
Q_{wu} = current STP flow (assuming 50% utilisation, 5,000 m^3/d);
Q_s = septage flow (60.8 m^3/d);
B_w = concentration of BOD in the sewage (250 mg/l);
B_s = BOD concentration in the liquid fraction of separated
    sludge (2,000 mg/l);
N_w = concentration of NH$_4$ in the sewage (50 mg/l);
N_s = NH$_4$ concentration in the liquid fraction of separated sludge
    (1,000 mg/l)
Effluent standards for BOD are assumed at 50 mg/l, and TKN
  at 10 mg/l
Oxygen required for BOD removal: 1.5 kg O$_2$/kg BOD
  removed
Oxygen required for TKN nitrification: 4.57 kg O$_2$/kg TKN nitrified
Cf = correction factor to account for lower oxygen transfer
  (0.7–0.95)
Effluent standard for BOD assumed: 50 mg/l
Effluent standard for NH$_4$ assumed: 10 mg/l

Oxygen required under design condition:

$$= 1.5 \times Q_{wd}(B_w - 50) + 4.57 \times Q_{wd}(N_w - 10)$$
$$= (1.5 \times 10{,}000 \times (250 - 50) + 4.57$$
$$\times 10{,}000 \times (50 - 10))/1{,}000$$
$$= 4{,}828 \text{ kg}$$

Co-treatment condition (the oxygen required will be adjusted, dividing by the correction factor)

$$= \frac{1.5 \times (Q_{wu}(B_w - 50) + Q_s(B_s - 50)) + 4.57 \times (Q_{wu}(N_w - 10) + Q_s(N_s - 10))}{0.7 \times 1{,}000}$$

$$= \frac{1.5 \times (5{,}000\,(250 - 50) + 60.8\,(2{,}000 - 50)) + 4.57 \times (5{,}000\,(50 - 10) + 60.8\,(1{,}000 - 10))}{0.7 \times 1{,}000}$$

$$= 4{,}096 \text{ kg}$$

> This is still below design conditions. Oxygen is likely to be a critical limiting factor for higher percentage of sludge added, where sludge BOD is high, STP current utilisation is high. With solids/liquid separation, the impact can be greatly mitigated.
> *See Excel sheet 3.*

graphs are based on conservative assumptions, and the actual conditions may allow slightly higher loadings.

10.4 Limit due to non-biodegradable fractions

The concentrations of soluble non-biodegradable fractions in sludge are unaffected by biological treatment processes, and will be retained in the effluent. When combined with sewage, these soluble non-biodegradable components will be diluted. This resulting concentration will set a limit for the allowable sludge volumes that can be co-treated.

If the soluble non-biodegradable COD fraction is known, then it can be used to estimate the resultant COD of the final effluent due to this fraction. Where this soluble non-biodegradable fraction of the sludge COD constitutes a large part of the effluent COD, compliance to COD standard becomes difficult. Limitations will be due to availability of oxygen, reactor residence time and the biodegradability of the sewage portion itself.

In the absence of reliable figures for COD fractionalisation, indicative figures available in literature may be assumed (see Figure 11). However, verification from local field tests should be carried out for confirmation when possible.

The graph in Figure 12 shows the soluble non-biodegradable COD as a percentage of final effluent COD, for different percentages of sludge co-treated, and different percentages of soluble non-biodegradable COD in sludge. When the soluble non-biodegradable COD as a percentage of final effluent COD

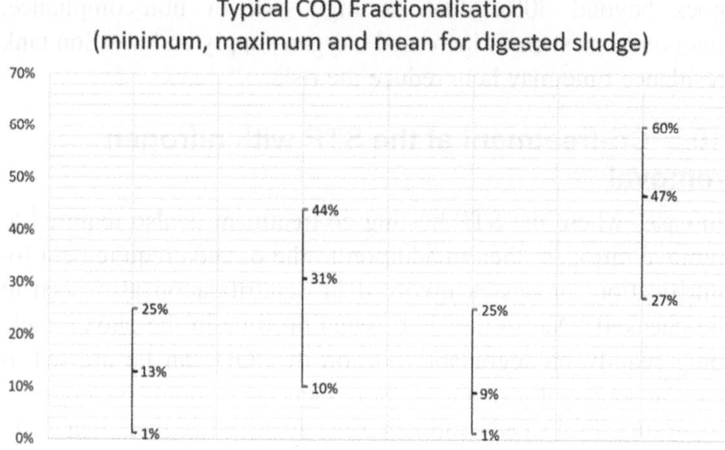

Figure 11 Typical COD fractionalisation. *Source*: Data from Tayler (2018).

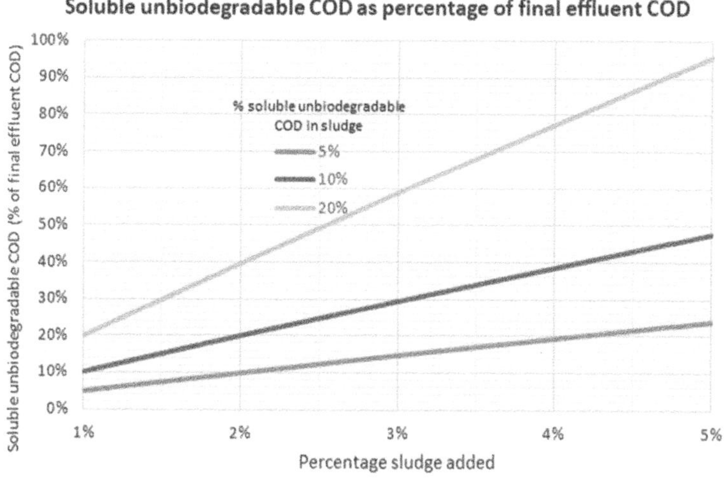

Figure 12 Soluble non-biodegradable COD in final effluent. COD: chemical oxygen demand.

goes beyond 30%, there is high risk of non-compliance. Interventions such as increased oxygen supply and aeration tank residence time may help reduce the risk.

10.5 Co-treatment at the STP with nitrogen removal

In cases where the STP hosting co-treatment is also required to remove nitrogen, then in addition to the oxygen requirement for nitrification, processes involved in denitrification also need to be checked. Due to the short retention time in the anoxic tank, only readily biodegradable portion of COD can be utilised as carbon sources for denitrification. General recommendations are to ensure BOD:TKN ratio in sewage to be higher than 3, to obtain good nitrification.

The sludge/sewage combined BOD:TKN ratio should be checked, and could be a limiting condition for the percentage sludge addition.

11 CO-TREATMENT IN OTHER SYSTEMS
11.1 Sequential batch reactors

Sequencing batch reactors (SBRs) are essentially activated sludge systems that are operated in batch mode. Aeration and settling take place in the same tank over different time sequences, rather than in separate tanks as in conventional activated sludge systems. The usual sequence is Fill – Aerate – Settle – Decant. There are a few variations of the SBRs, with the sequence being varied accordingly.

Typical SBR plant consists of a minimum of two reactors in a plant. When one unit of the reactors is in fill mode, the other reactors may be in the stage of react, settle, decant or idle. Continuous fill and intermittently decant system is one of the variations of this system, where feeding into all rectors are continuous but the other phases (react, settle, decant, idle) are run in sequence. In the reaction stage, oxygen supplied to the system shall be in accordance to the load to the system within the time frame of reaction cycle. This generally requires higher oxygen capacity per unit time than a continuously aerated system. In the decant stage, there shall be sufficient time to allow for MLSS to settle before effluent decanting begins. Decanting time is normally much shorter than fill time.

All SBR plants must be designed to cater for peak flows. Alternatively, an equalisation tank can be provided.

SBR processes are often controlled by programmable logic controllers (PLCs), with input parameters measuring the strength and volume of flow, and dissolved oxygen levels. This provides flexibility in operating to cater for variations in flow and load. As in other activated sludge processes, the solids concentration in the reactor (MLSS), food to micro-organisms ratio, mean cell residence time, sludge return rate and sludge wasting rate are parameters controlled to maintain the balance in the reactor.

The unit processes of the SBR and conventional activated sludge systems are the same. However, due to batch operation, the wastewater enters a partially filled reactor, containing biomass, which is acclimated to the wastewater constituents during preceding cycles. Once the reactor is full, it behaves like a conventional activated sludge system, but without a continuous influent or effluent flow. The aeration and mixing is discontinued after the biological reactions are complete, the biomass settles and the treated supernatant is removed. Excess biomass is wasted at any time during the cycle.

For co-treatment in SBRs, preliminary steps such as screening, grit removal and blending/mixing, and solids/liquid separation, and balancing tanks to blend the sludge and equalise it, by mixing into the sewage to avoid shock loads, should be provided as in other systems.

Where the incoming sludge is predominantly fresh, a stabilisation step may be appropriate before solid liquid separation or as part of solids–liquid separation.

For the SBR, the likely impacts from the supernatant on liquid stream processes are:

(1)   Insufficient aeration capacity for combined sewage/sludge load
(2)   Lower oxygen transfer rate and uptake due to higher solids in reactor (both dissolved and suspended)
(3)   Lower mixing effect due to high solids content
(4)   Higher solids loading and residence time in reactor
(5)   Higher solids overflow, considering possible lower settleability
(6)   Higher oxygen uptake for nitrification
(7)   Unavailability of sufficient organic carbon for denitrification

SBR processes have short retention times, and therefore instantaneous flows or at least hourly peak flows should be checked.

In the event there is no equalisation step, sewage flow variation over the day (and corresponding SS variation) will coincide with sludge flow based on tanker arrival and discharge rates. There will be a shock increase of solids during the period of sludge discharge. Process control should be carefully managed to adjust sludge wasting rates and reactor MLSS to maintain stable conditions during this period. This is particularly so, because in the batch operation of the SBR, the wastewater enters a partially filled reactor already containing biomass, which is acclimated to the wastewater constituents during preceding cycles. If the subsequent wastewater characteristics are substantially different, shocks happen.

Other considerations for co-treatment similar to extended aeration systems (as detailed in previous sections) also apply.

11.2 Ponds

Sewage stabilisation ponds are often used in series of three types of ponds:

(1) Anaerobic ponds for settling of SS and subsequent anaerobic digestion
(2) Facultative ponds for the remaining SS to settle, with aerobic stabilisation of dissolved organics at the surface, and anaerobic conditions at the bottom
(3) Maturation ponds for disinfection and stabilisation

Design criteria used are empirical, based on volumetric organic loading (organic load per unit pond volume) for the anaerobic ponds, and surface organic loading (organic load per unit surface area of pond) for facultative ponds.

Anaerobic ponds are 2–3 m deep, and typically remove 70–75% of BOD in tropical high temperature conditions when loaded with 250–350 g $BOD/m^3/day$.

Facultative ponds are 1–2 m deep and when loaded with 350 kg BOD/ha/day, remove a further 70% of BOD. Usually

Table 7 Pond loading rates.

	Design Condition	Co-treatment Condition
BOD loading rate to the anaerobic pond	$\dfrac{Q_{wd} \times B_w}{\text{Volume of anaerobic pond}}$	$\dfrac{(Q_{wu} \times B_w + Q_s \times B_s)}{\text{Volume of anaerobic pond}}$
BOD surface loading rate to the facultative pond	$\dfrac{(Q_{wd} \times B_w) \times (1 - B_r/100)}{\text{Surface area of facultative pond}}$	$\dfrac{(Q_{wu} \times B_w + Q_s \times B_s) \times (1 - B_r/100)}{\text{Surface area of facultative pond}}$

Q_{wd} = STP design sewage flow;
Q_{wu} = current STP sewage flow;
Q_s = septage flow;
B_w = concentration of BOD in the sewage;
B_s = BOD concentration in the septage/FS;
B_r = percentage BOD removed in anaerobic pond

Table 8 Volumetric load criteria for anaerobic ponds.

Temperature (°C)	Volumetric BOD Load (g BOD/m^3 d)
<10	100
10–20	$20T - 100$
20–25	$10T + 100$
>25	350

no mechanical aeration is provided, and oxygen is obtained by diffusion from the atmosphere, as well as from algae in the facultative ponds, which supply oxygen through photosynthetic process. The retention time of ponds is in the order of days, and this makes them good at handling variations in flow.

Direct addition of sludge to anaerobic ponds, for combined solids/liquid separation and the first stage in biological treatment may be considered where the sludge has a low solids content, preferably 1% or less.

As already explained in Section 9, the increase of peak flows for the combined flow is unlikely to be significant.

The organic loading under co-treatment conditions is compared with the design organic loading for the ponds to check adequacy (Table 7).

Mara (2004) recommends allowable volumetric organic loading rate based on ambient temperature as in Table 8 for anaerobic ponds.

For facultative ponds, Mara (1987, 2004) recommends surface organic loadings based on temperature as below:

Surface loading rate in kg BOD5/had

$$= 350(1.107 - 0.002T)^{T-25}$$

T is the mean temperature of the coldest month in °C.

With co-treatment, the limiting conditions can be caused by the high solids (especially if solids/liquid separation is not carried out ahead of the pond), which will settle out in the ponds, as well as by

---

BOX 13 POND LOADING

Anaerobic pond

Loading under design conditions:

$$\frac{Q_{wd} \times B_w}{\text{Volume of anaerobic pond}}$$

Loading under Co-treatment condition:

$$\frac{(Q_{wu} \times B_w + Q_s \times B_s)}{\text{Volume of anaerobic pond}}$$

Facultative pond

Loading under design conditions:

$$\frac{(Q_{wd} \times B_w) \times (1 - B_r/100)}{\text{Surface area of facultative pond}}$$

Loading under so-treatment condition:

$$\frac{(Q_{wu} \times B_w + Q_s \times B_s) \times (1 - Br/100)}{\text{Surface area of facultative pond}}$$

where:

Q_{wd} = pond design sewage flow (10,000 m^3/d);
Q_{wu} = current pond sewage flow (5,000 m^3/d);
Q_s = septage flow (60.8 m^3/d);
B_w = concentration of BOD in the sewage (250 mg/l);
B_s = BOD concentration in the septage/fecal sludge (2,000 mg/l);
B_r = percentage BOD removed in anaerobic pond (70%);

Volume of anaerobic pond: 15,000 m^3
Surface area of facultative pond: 2.5 ha
Temperature: 25°C

Anaerobic pond

Loading under design conditions

$$= \frac{10,000 \times 250}{15,000} = 167\,g/m^3 d$$

Loading under co-treatment condition

$$= \frac{(5{,}000 \times 250 + 60.8 \times 2{,}000)}{15{,}000} = 91\,g/m^3 d$$

This is well below design conditions as well as below the recommended value of 350 g/m^3 d

Facultative pond

Loading under design conditions

$$= \frac{(10{,}000 \times 250) \times (1 - 70/100)}{2.5 \times 1{,}000} = 300\,kg/ha\,d$$

Loading under co-treatment condition

$$= \frac{(5{,}000 \times 250 + 60.8 \times 2{,}000) \times (1 - 70/100)}{2.5 \times 1{,}000}$$

$$= 165\,kg/ha\,d$$

This is still below design conditions as well as below the recommended value of 350 kg/ha d
See Excel sheet 5.

ammonia, which would inhibit the process, use up available oxygen and interfere with algal growth.

The high solids content of sludge will result in the ponds filling up faster, and desludging of the ponds would be required at more frequent intervals. With co-treatment of well-digested sludges, a large portion of the solids settling out are already stabilised, and will not undergo further digestion and volume reduction. This will accelerate sludge accumulation in the pond. Desludging of ponds is a costly process which disrupts operations, especially if the ponds operate in a single stream (if duplicate pond streams are provided, one stream can be shut down during desludging).

Ponds should be desludged when sludge accumulation has reached 20–25% of the pond volume. If sludge removal is neglected, the performance of a pond will deteriorate and it will eventually fail. The desludging interval for anaerobic ponds that treat municipal sewage is typically measured in years, but the high solids content of co-treated sludge means that the desludging interval for ponds that are not preceded by other forms of solids/liquid separation is likely to be in months. Standard operating procedures should include guidance on monitoring the sludge accumulation rate in the ponds.

Solids liquid separation ahead of co-treatment in the ponds will reduce the solids loading on the ponds, and alleviate this problem.

BOX 14 POND DESLUDGING

The anaerobic pond will accumulate solids and would need to be desludged when one third full.

Daily dry solids loading under co-treatment condition:
$$Q_{wu} \times S_w + Q_s \times S_s \text{ kg/day}$$

A part of these solids will be settled in the anaerobic pond (S_{ret})
A part of the settled solids will be destroyed through anaerobic digestion (S_{des})

Daily dry solids accumulated: $(Q_{wu} \times S_w + Q_s \times S_s) \times S_{ret}$
$$\times (1 - S_{des}/100)$$

The volume of the wet sludge will depend on the solids content, S_c %

Wet sludge volume $= [(Q_{wu} \times S_w + Q_s \times S_s) \times S_{ret}$
$$\times (1 - S_{des}/100)]/(S_c/100)$$

Q_{wu} = current pond sewage flow (5,000 m^3/d);
Q_s = sludge flow (60.8 m^3/d);
S_w = suspended solids concentration in the sewage (300 mg/l);
S_s = suspended solids concentration in the sludge (10,000 mg/l);
S_{ret} = percentage solids settled in anaerobic pond (55%);

S_{red} = percentage solids destroyed in anaerobic pond (20%);
S_c = wet sludge solids content (10%);
V_p = volume of anaerobic pond: 15,000 m^3

Wet sludge volume = [(5,000 × 300 + 60.8 × 10,000)/

$$1,000 × 55\% × (1 - 20/100)]/$$

$$(10/100)\,\text{litres}$$

$$= 9.275\,\text{m}^3\,\text{wet sludge/day}$$

Assuming desludging when one third full of sludge,

Period before desludging: $(V_p/3)/$(daily wet sludge volume)

$$= 15,000/3/9.275\,\text{days}$$

$$= 539\,\text{days}$$

Similar calculation for the design condition of the pond (without co-treatment) may be done to compare the increase in frequency of desludging required.
See Excel sheet 5.

11.3 Upflow anaerobic sludge blanket reactors

Upflow anaerobic sludge blanket (UASB) reactors use an anaerobic process. The design criteria for UASBs include organic loading rate (OLR) and upflow velocity. The feed rates to the reactors must be consistent and uniform in order to maintain an effective sludge blanket for varying flows.

Successful operation of UASBs depends on regular monitoring of sludge levels and SS concentrations and withdrawal of excess sludge from the reactor.

Sludge retention time should be between 32 and 45 days while volumetric loading rate should be between 1.15 and 1.45 kg-COD/m^3 d (van Lier *et al.*, 2010).

For co-treatment, preliminary steps such as screening, grit removal and blending/mixing, and solids/liquid separation, and balancing tanks to blend the sludge and equalise it, by mixing into the sewage to avoid shock loads, should be provided as in other systems.

For the UASB, the likely impacts from the supernatant on liquid stream processes are:

(1) Insufficient sludge retention time for combined sewage/sludge load
(2) Volumetric organic loading rate exceeded

Sludge retention time and volumetric loading rate should be checked under co-treatment conditions to ensure design conditions are not exceeded.

Shock increase of solids during the period of sludge discharge must be anticipated and process control carefully managed to adjust rate feed rates and withdrawal of excess sludge from the reactor.

12 PRACTICAL CONSIDERATIONS

Commonly cities would encounter complex scenarios with some fresh and some digested sludge to be managed. The decision on the approach may be made after a study of the respective proportions or a characterisation analysis. Settleability tests (simple SVI tests) will be useful as well.

The proportions of fresh/digested sludge could also vary seasonally, and suitable approaches should be taken to deal with these situations.

There will be situations when a few tankers arrive bringing fresh sludge, and then a few bring digested sludge. Blending may help solve this, or separate streams for fresh and digested sludge may be justified if this is a common recurring condition.

13 OTHER CAUTIONARY FACTORS

(a)   The co-treatment should be commenced with an initial conservative loading limit, which may be varied after actual monitoring.

(b)   STPs selected for co-treatment should be reasonably large to justify the proper reception facilities necessary, and ensure operators are sufficiently skilled to manage the co-treatment. A minimum capacity of 5 million litres per day is suggested.

(c)   Information on the STP design basis and operating conditions should be obtained wherever possible, and this should then enable a good analysis of design vs actual conditions, and a good indicator of the potential of co-treatment at the STP.

(d)   STPs hosting co-treatment should also have sufficient spare capacity which is unlikely to be taken up by increased sewage flows in the planning period. A minimum spare and available capacity of 30% is suggested.

(e)   Where end products are required to conform to other parameters such as fecal coliforms, nitrogen, phosphorus, helminths, etc, appropriate processes shall be provided in the STPs concerned and the impact of the co-treatment on these processes shall also be assessed.

(f)   The STP should be close enough for easy access by tankers transporting the sludge. At the same time, there should be sufficient buffer distance so that nuisances to neighbourhoods can be minimised. The tanker routes to the STP should also avoid residential areas.

(g)   Where logistics warrant it, properly designed decanting stations may be set up to allow controlled sludge addition to sewers/pump stations. Proper reception facilities such as grit removal, screening, solids/liquid separation should be provided. Mechanisms to monitor sludge discharges should be in place at these locations.

(h) Crude sludge addition to sewers/pump stations or STP inlets should be the exception rather than the norm. It may only be considered when very small quantities of sludge are added to large STPs, usually as an interim arrangement before proper reception facilities for co-treatment arrangements are provided (grit removal, screening, solids/liquid separation, etc). Indiscriminate dumping of sludge in STPs or sewers will cause major damage to the systems and must be discouraged.

(i) Cities planning new STPs should consider the reality that parts of the city or its adjacent areas will continue to use on-site systems, and co-treatment of the sludge should be considered in the planning and design stage of the STP itself. This is an opportunity which has huge potential in countries like India where many STPs are being planned for implementation in the short to medium term.

(j) For planning purposes, this should be projected, considering better enforcement, greater success rate, containment retrofits, etc. The ultimate goal shall be emptying frequencies that are appropriate for the local context, considering the most suitable containment, ground water table, contamination risk, sludge accumulation rate for the type of containment used, etc.

(k) The variability of sludge entering the system can cause process upsets due to shock loadings. Most STPs are biological reactors, and microbial populations needs to be built up. Shock loadings may cause imbalances if the microbial biomass does not have sufficient time to acclimatise. The intermittent addition of sludge should be mitigated as much as possible, through regulation of tankers, blending/holding tanks, flow mixing and balancing.

(l) In situations where the emptying operations and tanker operators are not well regulated, risks of toxic or extraneous matter would exist. More control should be introduced, with logs of sources of sludge, tracking of tankers using GPS, inspection and sampling, and with holding tanks as safety buffer.

(m)   The risk of contaminated sludge may also result in the end products of the treatment (effluent and bio-solids) becoming toxic and unsuitable for reuse.

(n)   Control and monitoring mechanisms shall be put in place to ensure only domestic sludge (FS/septage) is brought in, and sludge is not contaminated by trade or industrial wastes. Visual inspection (any colour of sludge other than brownish/black), different or chemical odour can be used to identify industrial waste in septage. Operators can readily check for pH, conductivity, odour and colour to identify loads that contain commercial or industrial chemicals. When there is doubt, laboratory tests may be performed to identify the presence of industrial chemicals and heavy metals in septage.

(o)   Tanker operators should be regulated. Awareness, training, SOPs, monitoring and enforcement on private desludging operators will help to control the sludge management as a whole.

(p)   Institutional issues: if the STP operating entity is not also responsible for on-site sludge management, there may be an unwillingness to accept sludge for co-treatment. A suitable institutional arrangement or regulatory mechanism would be needed.

(q)   Finally, this is not a comprehensive design manual, and is intended to guide the planner and in preliminary design stage to assess the option of co-treatment. Where co-treatment is to be implemented, detailed design considerations shall be carried out before implementation. The considerations in this guide have adopted a very conservative approach in suggesting low risk levels of co-treatments, and with more detailed data and knowledge of the local conditions, higher levels are probably possible.

CONCLUSION

Co-treatment (even initially crude sludge co-treatment, and later liquid portion after solid/liquid separation) is very promising as a starting strategy for many cities with existing large STPs with sufficient capacity. It is a quickly implementable solution, and where the current practise is open dumping of the sludge, it will result in a huge improvement. It may be adopted as an interim arrangement and stepped up gradually. It could be a part of an overall sludge and sewerage strategy which would strategise utilisation of available facilities for sewage and sludge treatment in the most optimum manner, considering capacities, risk, logistics and planning timeframe.

Moreover, where new sewage treatment facilities are being planned, the planning and design should take into account the reality that some of the urban areas in the proximity of the STP will continue to use on-site systems in the near future. The sludge from these areas is best co-treated in the nearby STPS, and the design of these STPS should therefore accommodate this.

All this must be coupled with good regulation and control of FS emptying and transport to mitigate risks as mentioned above. Of course, good operational control of STP process is crucial too. Eventually as dedicated sludge facilities come on board, the practice can be slowly phased out in the more risky cases.

We should also bear in mind the likelihood of future flows of sewage (likely to increase due to growth of population, increased coverage, increased connections, etc) and sludge (likely to increase with growth of population, eradication of open defecation, better toilets and on-site systems, scheduled emptying etc) and the timeframe of these projected flows.

Many variations of the strategy are possible, from co-treating both liquid and solid portions, or just the liquid, to co-siting sludge treatment facilities at STPs.

It is hoped that this guideline makes the issues clearer, affords a better understanding of the solution options and risks and enables planners and decision makers to avail of the available options through co-treatment.

APPENDIX: USING THE WORKSHEETS

There are five worksheets. The first three are focused on specific treatment stages, to enable better understanding of the impact of co-treatment on these process stages.

The last two are more complete worksheets for the whole treatment process, for activated sludge and for oxidation ponds.

All the sheets are protected except for cells where data entry by user is required (blue cells). There is no password and user may unprotect the sheet and view/modify formulas and other data.

(1)   **Sheet 1: This is focused on solids/liquid separation.**
       Input data include:
       (a)   STP design flow (m^3/d)
       (b)   STP current utilisation (%)
       (c)   Sludge added (daily)
       (d)   Working days/hours for tanker reception
       (e)   Sludge solids concentration
       (f)   Solids/liquid separation: % solids removal
       (g)   Expected solids content of solid stream of sludge

       Calculated:
       (a)   Total solids (kg) in incoming sludge,
       (b)   Solids in solids stream (kg)
       (c)   Solids in liquid stream (kg)
       (d)   Solids stream flow (m^3/d)
       (e)   Liquid stream flow (m^3/d)

(2)   **Sheet 2: This is modelling primary settling tank.**
       Input data
       (a)   STP design flow (m^3/d)
       (b)   STP current utilisation (%)
       (c)   Sewage solids concentration
       (d)   Primary settling tank: % solids removal
       (e)   Sludge added (daily)
       (f)   Working days/hours for tanker reception
       (g)   Sludge solids concentration

Calculated:

(a) Primary settling tank percentage increase in daily & hourly solids

(b) Primary settling tank increase in hourly peak flow

(3) **Sheet 3: This is modelling the aeration tank.**
Input data
(a) STP design flow (m^3/d)
(b) STP current utilisation (%)
(c) Sewage concentration:
  • SS
  • BOD
  • TKN
(d) Sludge added (daily)
(e) Working days/hours for tanker reception
(f) Sludge concentration
  • SS
  • BOD
  • TKN
(g) Solids/liquid separation: % removal of:
  • SS
  • BOD
(h) Expected solids content of solid stream of sludge
(i) Effluent standards:
  • BOD
  • TKN
(j) Oxygen required for:
  • BOD reduction (kg O_2/kg BOD)
  • TKN reduction (kg O_2/kg TKN)

Note:

  • *If there is no solid/liquid separation provided, the values for solids/liquid separation % removal of SS and BOD can be set to zero.*

Calculated:

(a) Solids loading (design condition & co-treatment condition)

(b) Percentage increase in solids (daily) due to co-treatment

(c) Oxygen required (design condition & co-treatment condition)

(d) Percentage increase in oxygen required due to co-treatment

(4) **Sheet 4: This models an activated sludge treatment plant.**

Input data

(a) STP design flow (m^3/d)

(b) STP current utilisation (%)

(c) Sewage concentration:

- SS
- BOD
- TKN

(d) Sludge added (daily)

(e) Working days/hours for tanker reception

(f) Sludge concentration

- SS
- BOD
- TKN

(g) Solids/liquid separation: % removal of:

- SS
- BOD

(h) Expected solids content of solid stream of sludge

(i) Primary settling tank: % removal of:

- TSS
- BOD
- TKN

(j) Effluent standards:

- BOD
- TKN

(k)   Oxygen required for:
- BOD reduction (kg O_2/kg BOD)
- TKN reduction (kg O_2/kg TKN)

Notes:

- *If there is no solid/liquid separation provided, the values for solids/liquid separation % removal of SS and BOD can be set to zero.*
- *If there is no primary settling tank provided, the values for primary tank % removal of SS, BOD and TKN can be set to zero.*

Calculated:

(a)   Primary settling tank:
(l)percentage increase in daily & hourly solids
(m)percentage increase in hourly peak flow

(b)   Aeration tank:
- solids loading (design condition & co-treatment condition)
- percentage increase in solids (daily) due to co-treatment
- oxygen required (design condition & co-treatment condition)
- percentage increase in oxygen required due to co-treatment

(5)   **Sheet 5: This models a pond system.**
Input data

(a)   Ambient temperature (degrees C)
(b)   Pond design flow (m^3/d)
(c)   STP current utilisation (%)
(d)   Sewage concentration:
- SS
- BOD

(e)   Sludge added (daily)
(f)   Working days/hours for tanker reception

(g) Sludge concentration
  - SS
  - BOD
(h) Solids/liquid separation: % removal of:
  - SS
  - BOD
(i) Expected solids content of solid stream of sludge
(j) Anaerobic pond
  - Volume
  - expected BOD reduction
  - expected solids settled
  - expected solids reduction (due to anaerobic digestion)
(k) Facultative pond surface area

Notes:

- *If there is no solid/liquid separation provided, the values for solids/liquid separation % removal of SS and BOD can be set to zero.*

Calculated:

(a) Anaerobic pond volumetric loading (design condition and co-treatment condition)
(b) Maximum recommended Anaerobic pond volumetric loading
(c) Facultative pond surface loading (design condition and co-treatment condition)
(d) Maximum recommended facultative pond surface loading
(e) Estimated desludging frequency (design condition and co-treatment condition)

ACKNOWLEDGMENTS

I would like to thank the following people (in no particular order) for their kind efforts in reading, reviewing and providing most useful critical feedback, comments, suggestions and encouragement, without which this guide would not be what it is.

Ravikumar Joseph, Doulaye Kone, K. E.Seetharam, Pierre Flamand, Michael McWhirter, Roshan Shrestha, Chengyan Zhang, Ambarish Karunanithi, Krishna C. Rao, Vrishali Subramaniam and Isha Basyal.

All credit for the usefulness of the guide goes to these people, while I take responsibility for any inaccuracies and other shortcomings.

In line with the intention of the guide to help planners, practitioners and others to get some clarity to be able to proceed with considering co-treatment options, I welcome feedback and suggestions, so that future revisions of this document may be more useful.

Please email to me at dorainarayana@gmail.com.
Dorai Narayana

REFERENCES

Mara D. D. (1987). Waste stabilization ponds: problems and controversies. *Water Quality International*, **1**, 20–22.

Mara D. D. (2004). Domestic Wastewater Treatment in Developing Countries. Earthscan, London.

Metcalf & Eddy, Inc. (1991). Wastewater Engineering: Treatment, Disposal, Reuse. McGraw-Hill, New York.

Tayler K. (2018). Fecal Sludge and Septage Treatment: A Guide for Low- and Middle-Income Countries. Practical Action Publishing Ltd, Rugby, UK.

Van Lier J. B., Vashi A., Van Der Lubbe J. and Heffernan B. (2010). Anaerobic sewage treatment using UASB reactors: engineering and operational aspects. In: Environmental Anaerobic Technology, H. H. P. Fang (ed.), Imperial College Press, London, pp. 59–89.

FURTHER READING

Gupta S., Jain S. and Chhabra S. S. (2018). Draft Guidance Note on Co-Treatment of Septage at Sewage Treatment Plants in India, April 2018. https://www.fsmtoolbox.com/assets/pdf/150._Guidance_Note_on_ Co-treatment_April_2018.pdf. [Accessed 23 Jan 2020]

Malaysian Sewerage Industry Guidelines. SPAN website http://www.span. gov.my

Robbins D., Sher Singh P. E. and Kearton R. (2017). Co-treatment of Septage with Municipal Wastewater in Medium Sized Cities in Vietnam. Paper Presented at FSM4, 2017, Chennai, India.

IWA Publishing's authorised EU representative for General Product
Safety Regulations is Diane D'Arras, 15 rue Duret, 75116 Paris,
France, e-mail: safety@iwap.co.uk.

Printed and bound by CPI Group (UK) Ltd, Croydon, CR0 4YY
06/07/2026
02157563-0001